THE
SELF-ACTUALIZING
COSMOS

"In his new book, Ervin Laszlo, the world's foremost systems theorist, summarizes his evolutionary connectivity hypothesis and his concept of the Akashic field. He offers a brilliant solution for the paradoxes and anomalous phenomena that emerged in the course of the twentieth century in a broad variety of fields—from astrophysics, quantum-relativistic physics, and chemistry to biology, anthropology, thanatology, parapsychology, and psychology. In this tour de force, Laszlo's renaissance intellect spanning a broad range of scientific disciplines offers the reader a breathtaking vision of a self-actualizing cosmos. A book that should not be missing in the library of any scientist seeking to understand the nature of reality."

STANISLAV GROF, M.D., AUTHOR OF *HOLOTROPIC BREATHWORK*, *PSYCHOLOGY OF THE FUTURE*, AND *WHEN THE IMPOSSIBLE HAPPENS*

"The modern idea of a mindless, purposeless, directionless universe will take its place as a morbid relic in the history of ideas. It is being replaced by the vision expressed in *The Self-Actualizing Cosmos*. This important book restores hope and meaning and shows that the cosmos is a friendlier home than we have recently taken it to be."

LARRY DOSSEY, M.D., AUTHOR OF *ONE MIND: HOW OUR INDIVIDUAL MIND IS PART OF A GREATER CONSCIOUSNESS AND WHY IT MATTERS*

"Forty years ago Ervin Laszlo changed the way we saw the world with his classic book *The Systems View of the World.* Today he changes the way we understand the cosmos and our own place in it. This is a groundbreaking and very readable work."

ALLAN COMBS, PH.D., AUTHOR OF
CONSCIOUSNESS EXPLAINED BETTER, SYNCHRONICITY,
AND *RADIANCE OF BEING* AND DIRECTOR OF THE
CALIFORNIA INSTITUTE OF INTEGRAL STUDIES (CIIS)
CENTER FOR CONSCIOUSNESS STUDIES

THE
SELF-ACTUALIZING
COSMOS

The AKASHA REVOLUTION in SCIENCE and HUMAN CONSCIOUSNESS

ERVIN LASZLO

Inner Traditions
Rochester, Vermont • Toronto, Canada

Inner Traditions
One Park Street
Rochester, Vermont 05767
www.InnerTraditions.com

Text stock is SFI certified

Library of Congress Cataloging-in-Publication Data

Laszlo, Ervin, 1932– author.
  The self-actualizing cosmos : the Akasha revolution in science / Ervin Laszlo.
    pages cm
  Includes bibliographical references and index.
  Summary: "An exploration of the current revolution in scientific thought and the
newest scientific findings in support of the Akashic field"—Provided by publisher.
  ISBN 978-1-62055-276-6 (pbk.) — ISBN 978-1-62055-277-3 (e-book)
  1. Cosmology. 2. Consciousness. I. Title.
  BD511.L377 2014
  501—dc23
                                                                    2013031097

Printed and bound in the United States by Lake Book Manufacturing, Inc.
The text stock is SFI certified. The Sustainable Forestry Initiative® program
promotes sustainable forest management.

10 9 8 7 6 5 4 3 2 1

Text design and layout by Priscilla Baker
This book was typeset in Garamond Premier Pro with Swiss 721 used as a display
typeface

To send correspondence to the author of this book, mail a first-class letter to the
author c/o Inner Traditions • Bear & Company, One Park Street, Rochester, VT
05767, and we will forward the communication, or contact the author directly at
ervin@ervinlaszlo.it.

Contents

Acknowledgments

I would like to thank cutting-edge physicists Paul LaViolette and Peter Jakubowski for contributing the pioneering hypotheses that lay the foundations of the Akasha paradigm in physics.

I wish to express my thanks and appreciation to Edgar Mitchell, Stanley Krippner, David Loye, Kingsley Dennis, David Lorimer, Deepak Chopra, and Ken Wilber—long-standing friends and colleagues—for comments and suggestions that helped me achieve a comprehensive formulation of the science and philosophy of the Akasha paradigm.

I likewise acknowledge the contribution of Dr. Maria Sági, another long-standing friend and colleague, whose discoveries regarding space and time-independent nonlocal healing have motivated my development of the Akasha paradigm and have also kept me in good health for nearly three decades.

My sincere thanks to David William Gibbons and Györgyi Szabo for conversations that helped me spell out for the reader basic questions about the meaning and relevance of the new paradigm.

I am grateful to Marco Antonio Galvan for his deep interest in this paradigm and his remarkable acumen in finding and calling to my attention the cutting-edge scientific findings that testify to its scientific foundations.

It gives me particular pleasure to acknowledge the expert contribution of my sons Christopher and Alexander in bringing to my attention findings and ideas relevant to the exposition and implications of this paradigm.

Last but by no means least, I thank my wife Carita Marjorie, without whose unfailing patience and constant love and support I could not have had the endurance, the inspiration, and the concentration to work on the concepts and ideas I have attempted to express in this, my latest and perhaps most definitive "Akashic field" book.

Prologue

There is a major revolution under way in science today, a transformation that is both profound and fascinating. It changes our view of the world, and our concept of life and consciousness in the world. It comes at a propitious time.

We know that the world we have created is unsustainable: we need new thinking to avert a collapse and set us on course for a sustainable and thriving society. The inspiration for the new thinking can come from science but not, or not only, from science as a source of new technologies. Rather, we need to view science as a source of orientation and guidance, as a wellspring of trustworthy ideas for rediscovering our relations to each other and the universe. The revolution under way in science offers a paradigm that can fill this need.

A *paradigm* in science is the sometimes tacit but always effective foundation of the way scientists conceive of the world, including the objects and processes they investigate. A new paradigm is an important innovation in science: it allows scientists to piece together the emerging elements of scientific knowledge and perceive the meaningful whole that underlies this complex mosaic of data, theory, and application.

A new paradigm has meaning and interest well beyond science. It provides a holistic, integral view of life and universe, lifting these vistas from the realm of speculation into the domain of careful observation and rigorous reasoning. Although based on sophisticated theories

1

and wide-ranging observations, the now emerging paradigm is basically simple and inherently meaningful.

The book in the hands of the reader is dedicated to conveying the principles of the new paradigm in language that, to paraphrase Einstein, is as simple as possible—but not simpler. It outlines the principles of the new paradigm and applies them to enhance our understanding of cosmos and consciousness. It then turns to such matters of human concern as how we perceive the world; how we can use the information we receive from the world to maintain our health; what kind and level of freedom we enjoy in the world, and, last but not least, how we can strive for the highest value in life, which philosophers called "The Good."

The author has been engaged in exploring and elaborating the paradigm coming to light in science for more than four decades. He hopes that this study, the latest and most mature fruit of his endeavors, will live up to the expectation his friends and readers attach to it. And that it will prove to be an effective scheme for investigating and elaborating the new paradigm emerging in science in order that we may reach a better understanding of who we are, what the world is, and what our mission is in this crucial epoch of our history.

PART ONE

Conceptual Foundations of the New Paradigm

The paradigm emerging in the welter of observation, exploration, and debate in today's revolutionary period in science is a pathbreaking innovation, but not an ad hoc novelty. The new paradigm is solidly based on what science and scientists already know about the nature of reality: it recognizes the validity of the accumulated storehouses of scientific knowledge. But the new paradigm pieces together the elements of scientific knowledge in a fashion that is more consistent, coherent, and meaningful than that possible in the light of the old and still influential paradigm. It offers a new gestalt, a new way of organizing the dots of scientific knowledge, connecting them with optimum simplicity and coherence. And it does so with a measure of the elegance that scientists and philosophers have always sought in their theories.

Part one presents the conceptual foundations of the new paradigm prior to the exploration, in the subsequent parts, of its implications for our knowledge of the world and for our thinking and acting in the world.

1

Revolution in Science

Einstein remarked, "We are seeking the simplest possible scheme of thought that can tie together the observed facts." This phrase encapsulates the quintessence of the project known as science. Science is not technology: it is understanding. When our understanding of the world matches the nature of the world, we discover more and more of the world and have more and more ability to cope with it. Understanding is basic.

Genuine science seeks the scheme that could convey comprehensive, consistent, and optimally simple understanding of the world and ourselves in it. That scheme is not established once and for all; it needs to be periodically updated. The observed facts grow with time and become more diverse. Tying them together in a simple yet comprehensive scheme calls for revising and occasionally reinventing that scheme.

In recent years the repertory of observed facts has grown and has become highly diverse. We need a new scheme: a more adequate paradigm. This, in the language of Thomas Kuhn, means a new scientific revolution. In his seminal opus *The Structure of Scientific Revolutions*, Kuhn (1962) noted that science grows through the alternation of two radically different phases. There is the relatively enduring phase of "normal science," and there is the phase of "scientific revolutions." Normal science treads water: it is only marginally innovative. It ties together the observed facts within an established and consensually validated scheme, and if it encounters observations that do not fit that scheme, it extends and adjusts that scheme.

This, however, is not always possible. If the attempt is not relinquished, the established scheme becomes unmanageably complex and opaque, as Ptolemaic astronomy did through the constant addition of epicycles to its basic cycles to account for the "anomalous" movement of the planets. When that critical point is reached in the growth of science, it is time to replace the established scheme. There is a need for a new paradigm that could ground the theories and interpret the observations to which they refer. The relatively calm phase of normal science comes to an end and gives way to the turbulence that hallmarks a period of scientific revolution.

In the natural sciences the turbulent phase of a revolution has already started. A number of unexpected, and—for the dominant paradigm—critically anomalous observations have come to light. They call for a basic paradigm shift: for a fundamental revolution that reinterprets science's most basic assumptions about the nature of cosmos, life, and consciousness.

The series of critically anomalous observations can be traced to an experimental finding in the early 1980s. A paper by French physicists Alain Aspect and collaborators (Aspect et al. 1982) reported on an experiment carried out under rigorously controlled conditions. This experiment demonstrated that when particles are split and the split halves are projected a finite distance from each other, they remain connected despite the space that separates them. Moreover their connection is quasi-instant. This contradicts a basic tenet of relativity: according to Einstein's theory the speed of light is the highest speed at which anything or signal can propagate in the universe.

Aspect's experiment was repeated, and it always produced the same result. The science community was baffled but finally dismissed the phenomenon as without deeper significance: the "entanglement" of the split particles, physicists said, is strange, but it does not convey information or "act on" anything. But this, too, was placed in question in subsequent experiments. It turned out that the quantum state of particles, and even of whole atoms, can be instantly projected across any finite distance. This came to be known as "teleportation." Instant, quantum-resonance-based interactions have been discovered also in living systems, and even in the universe at large.

A related anomalous fact came to light in regard to the level and form of coherence found in complex systems. The observed coherence suggests instantaneous interaction between the parts or elements of the systems: interaction that transcends the recognized bounds of space and time. In the quantum realm, entanglement—instant connection among quanta (the smallest identifiable units of "matter") at any finite distance—has recently been observed not only across space but also across time. It has been known that quanta that at any one time had occupied the same quantum state remain instantly correlated; it now appears that quanta that had never coexisted at the same time (as one of the particles had ceased to exist before the other came into being) also remain instantly entangled.

This kind of entanglement is not limited to the quantum domain: it surfaces also at macroscopic scales. Life would not be possible in its absence. In the human body, for example, trillions of cells need to be fully and precisely correlated to maintain the organism in its physically highly improbable living state. This calls for quasi-instant multidimensional connection throughout the organism.

Yet another finding that cannot be explained by the current paradigm is that organic molecules are produced in stars. The received wisdom is that the universe is a physical system in which life is, if not an anomalous, at least an uncommon and very likely accidental phenomenon. After all, living systems can evolve only under conditions that are extremely rare in space and time. However, it turns out that the organic molecules on which life is based are produced already in the physical-chemical evolution of stars. The molecules are ejected into surrounding space; they coat asteroids and clumps of interstellar matter, including those that subsequently condense into stars and planets. It appears that the laws that govern existence and evolution in the universe are fine-tuned to produce the kind of complex systems we associate with the phenomena of life.

Observations of this kind cannot be accounted for by patching up the dominant paradigm: they challenge the very foundations of the basic scheme with which scientists have been tying together the observed facts. This was the case also at the turn of the twentieth century, when the science community shifted from the Newtonian to the relativity paradigm. It was the case in the 1920s as well, with a shift to the quantum paradigm. More limited paradigm shifts have unfolded in specific domains since then, with the emergence of transpersonal theories in psychology and the advent of non-Big Bang "multiverse" models in cosmology.

The paradigm emerging in science in the second decade of this century signifies a major shift in the worldview of science. It is a shift from the dominant paradigm of the twentieth century, where events and interactions were believed to take place in space and time and were considered local and separable, to a twenty-first century paradigm that recognizes that there is a deeper dimension beyond space and time and that the connection, coherence, and coevolution we observe in the manifest world are coded in the integral domain of that deeper dimension.

2

Fields

A world where connection, coherence, and coevolution are fundamental features is not a fragmented and fragmentable world, but an integral one. In this world nonlocality is a fundamental factor: things that occur at one place and time also occur at other places and times—in some sense, they occur at all places and times.

Nonlocality in the world is an inference from current observations, but it is not accounted for by the twentieth-century paradigm that still grounds those observations. There is an urgent need for a paradigm in which nonlocality is a basic feature—the paradigm of a world that is intrinsically nonlocal. Such a paradigm is now emerging at the leading edge of scientific inquiry. It is based on a new understanding of how parts interact within wholes; ultimately how the parts we know as quanta and the macroscale entities built as coordinated sequences of quanta interact within the largest whole we call "cosmos." The basic concept that can convey scientific meaning and legitimacy to this understanding is *field*.

CONCEPT OF FIELDS

Fields are bona fide elements of the physical world, although they are not in themselves observable. They are like fishing nets so fine that

8

their strands cannot be seen. Tugging at any link in the net nevertheless creates a corresponding movement in all the others.

The fields themselves are not visible, but they produce observable
effects. Fields connect phenomena. Local fields connect things within
a particular region of space and time, but there are also universal fields
that connect things throughout space and time. Quanta, and the things
constituted of quanta, interact through fields, and they interact univer-
sally. Universal fields mediate interaction throughout the universe, and
they mediate it nonlocally.

SCIENCE'S "CLASSICAL" FIELDS

The first fields postulated in science were needed to account for
attraction between things across space. Action at a distance was not
acceptable—even Einstein was not happy with the idea of events occurring at a distance without some form of connection between them: he
called it "spooky." Yet things attract each other across intervening space,
and classical physics came up with the concept of a field: the gravitational field.

In the early eighteenth century the gravitational field was assumed
to be built by mass-points in space and to act on each of the mass points
at its specific spatial location. Later the field concept was extended to
include electric and magnetic phenomena. In 1849 Michael Faraday
replaced direct action among electric charges and currents with electric and magnetic fields produced by all charges and currents at a given
time.

In 1864 James Clerk Maxwell went further: he proposed the electromagnetic theory of light. Here the electromagnetic (EM) field is universal: it accounts for electric and magnetic phenomena wherever they
occur. The observed phenomena are referred to as waves propagating
with finite velocity in the universal EM field.

By the dawn of the twentieth century physics had acquired four
universal fields, the long range gravitational and electromagnetic fields,

and the short range strong and weak nuclear fields. Since the middle of the last century these "classical" fields have been joined by a variety of nonclassical fields postulated in quantum field theory.

QUANTUM FIELDS

Quantum fields are complex entities: they describe phenomena in space and time, as well as spacetime itself. These phenomena are not material in the conventional sense of the term. Since the middle of the twentieth century there has not been anything in the world that, on a closer look, quantum physicists could identify as "matter." There were, and are, only fields in excited states, with the excitations appearing as material entities.

Both particles and forces are states of excitation of an underlying field. The universal forces are described as Yang-Mills fields, replacing the classical electromagnetic field.* Quanta, in turn, are described by what are known as *fermionic fields,* and the elusive particles that endow quanta with mass make up the *Higgs field,* an invisible energy field that exists throughout the universe. In the final count all physical phenomena are "field excitations," vibrational patterns in spacetime.

Space itself is not an independent variable in the field equations and is not considered an independent element in the universe. As described in string theory, the structure of space is directly dependent on the conditions that define the presence of the mass points classically known as matter. Spacetime as a whole is generated by fields.

According to quantum theory, at very small scales space is not smooth; it is not flat even in the absence of mass: it constitutes a turbulent "quantum foam." Mathematically cumbersome infinities are associated with the quantum foam (infinities in a theory claiming to describe a finite universe demonstrate a breakdown of its coherence), and string theory was developed to resolve these infinities. The theory eliminates

---

*Yang-Mills fields are named for quantum physicists Cheng Ning Yang and Robert Mills who put forward a theory of the behavior of elementary particles that led to the unification of the weak and the electromagnetic force.

them by "smearing" the short-distance properties of space, smoothing the quantum turbulence. In this context the elementary entities of the universe are vibrating filaments; they appear as particles because instrumentation cannot penetrate to the required scale. (Current technology allows measurements to 10^{-18} m, whereas the Planck scale required for string phenomena to appear is 10^{-35} m.) If this is the case, particles are epiphenomena created by the limitations of our system of observation.

Strings replace the massive particles that according to the general theory of relativity curve the four-dimensional matrix of spacetime. (The general theory of relativity is a geometric theory of gravitation put forward by Einstein in 1916. It provides a unified description of gravity as an intrinsic geometric property of the four-dimensional matrix called spacetime.) Electrons, muons, and quarks, as well as the entire class of bosons (light and force particles) and fermions (matter particles) are not particles but vibrational modalities defined in accordance with the geometry of spacetime. In the sophisticated forms of string theory spacetime is "stringy": the relative points of space are themselves superstrings. Empty space is a low vibrational pattern, a "hole" in Calabi-Yau space, and the phenomena classically seen as particles appear in the intersection of the boundaries of Calabi-Yau space-holes.

Although the theory of relativity as well as quantum field theory are highly accomplished schemes for understanding the connections that appear among phenomena in space and time, the fields they postulate do not offer an adequate explanation of the nonlocality that has been observed at the supersmall scale of the quantum and is now observed at macroscopic scales as well. It appears that a further item needs to be added to the repertory of fields known to science. It is to the nature of this "missing field" that we turn next.

3

The Connecting Holofield

Just as with attraction and repulsion among observed entities, and the transmission of force and light, the nonlocality coming to light in diverse domains of investigation calls for recognizing the action of a field, more specifically, the action of a "nonlocal interaction-generating" field. (An interaction is said to be nonlocal when it transcends the known limits of effect-propagation in space and time.) The concept of such a field cannot be an ad hoc postulate, nor can it be an extra-scientific hypothesis. It must be rooted in what science already knows about the nature of physical reality. The question we take up concerns the nature of such a field. There are theories in science that offer cogent starting points for tackling this question.

Nonlocal interaction can be referred to the conjugation of the waves emitted by quanta and systems of quanta. (Waves are said to be conjugate when their oscillations are synchronized at the same frequency.) The information present at the nodes of the interference patterns produced by conjugating waves is shared among the waves. Thus the synchronization of the phase of the waves emitted by quanta and systems of quanta correlates their state. This insight is basic for understanding the nonlocality of interactions in nature.

The problem is that phase relations are not readily observable. When we examine the physical properties of the systems that emit the waves, the waves emitted by them remain obscure; and when we focus on the waves, the physical properties of the systems become indistinct.

The relationship between the observation of the phase of the components of a system and the components themselves is analogous to the complementarity principle enunciated by Niels Bohr. Focusing on the atomic structure of the system under observation entails losing the dynamics of their wave frequencies; and focusing on the phase dynamics obscures the atomic structure. Yet if nonlocality in a system is due to the conjugation of the phase of the wave frequencies of its components, knowing the phase dynamics of the system is essential for understanding the origin of the nonlocality.

The complementary relationship between the observation of the components of a system and the observation of its phase dynamics was discovered in regard to liquid helium, a superfluid whose components are fully in phase and are extraordinarily coherent. Subsequently, unexpected forms and levels of coherence were discovered in macroscale systems also at everyday temperatures, including liquid water and living tissue. Classical quantum mechanics (QM) could not explain this phenomenon since it concentrated on the quantum components of the systems, and hence it failed to account for their phase dynamics. Quantum field theory (QFT) overcame this failing. In QFT the field that governs the phase of the system's components is as much a part of the system as the components themselves: there is no separation between components and their interaction.

The question that remains to be clarified is the nature of the waves involved in nonlocal interaction. Physicists had mostly maintained that these are electromagnetic (EM) waves. This is not satisfactory, however, as on macroscopic levels and with extended time frames, nonlocality in interaction calls for long-range phase conjugation. This cannot be attributed to electromagnetic waves since in the EM field the effect falls off with distance and time. Hence, if we are to account for nonlocality over extended time frames and distances, we must either redefine the properties of the EM field or recognize the presence of a different kind of field. Since EM theory is solidly established, it is more reasonable to inquire into the latter possibility.

This endeavor is promising. There is a kind of wave field that can explain nonlocal interaction at both micro- and macroscales, and over any finite distance: this is a field of scalar waves. Scalars are longitudinal rather than transverse waves such as EM waves, and they propagate at velocities proportional to the density of the medium in which they propagate. Their effect, unlike those of EM waves, does not fall off with distance and time.

Given these properties, it is plausible that the field responsible for nonlocal interaction in nature is a field of scalar waves. Since the propagation velocity of these waves is proportional to the density of the medium in which they propagate, and since space is known to be a superdense virtual energy medium, scalars can be expected to propagate in space at supraluminal (faster-than-light) velocities. Thus we can understand that the nonlocality of their interaction extends over vast distances.

PROPERTIES OF THE NONLOCALITY-GENERATING FIELD

We now consider the properties of the indicated nonlocal interaction generating field. Following the hypothetico-deductive method of theory-construction in science, these properties can be first "invented," but then they must be tested against observations. The properties can be considered verified when they provide the simplest consistent explanation of the observations.

We apply this principle to the field presumed to generate nonlocal interaction in nature. We ascribe the following properties to the field:

Universality (the field is present and active at all points in space and time)

Nonvectorial effectiveness (the field produces effect through nonvectorial in-formation)

Holographic information storage (information in the field is carried in a distributed form, with the totality of the information present at all points)

Supraluminal effect-propagation (the field produces effect quasi-instantly at all finite distances)

Effect-production through phase-conjugate resonance (the nonlocal effect is due to the conjugation of the waves of the field with those of the systems with which they interact)

We assume that the interaction of such a field with quanta and quanta-based systems—atoms, molecules, cells, organisms, ecologies, and systems even of cosmological dimension—produces nonlocal interaction within and among them.

The process whereby the field creates nonlocal interaction among quanta and quanta-based systems can be described as follows:

The scalar waves of the universal holofield interfere with the waves emanating from quanta and quanta-based systems, and the resulting phase-conjugating interference transfers information from the field to the systems. Since the field is universal and transmits information in the distributed mode of holograms, and the waves of the field are scalars that propagate quasi-instantly in space and do not attenuate in time, the transfer of information produces instant or quasi-instant interaction within and among quanta and quanta-based systems throughout the observable regions of space and time.

4

Fields, Physical Reality, and the Deep Dimension

The question we now raise concerns the physical reality of the nonlocal interaction-generating field. Is this field a bona fide element in nature? In tackling this question the first thing we need to do is to assess the reality of the fields known to science. Are these fields real elements of the world, or are they theoretical entities postulated to facilitate the comprehension of the real elements?

PHYSICS AND THE PROBLEM OF PHYSICAL REALITY

Asking whether we can ascribe physical reality to fields, the same as to any other entity in science, is a difficult question for scientists: it falls into the category of "quasimetaphysical" questions that are best left to philosophers.

Throughout the twentieth century theoretical scientists, in particular quantum physicists, preferred to deal with the "how" of phenomena and left aside questions of "what." They were reluctant to consider what quanta are "in themselves," contenting themselves with explaining their interactions. According to Nobel Laureate Eugene Wigner, physicists

16

should correlate observations and not deal with observables.

This was a useful strategy in the early days of quantum physics: it enabled exploratory work to proceed pragmatically, without the burden of requiring the theories to be concerned with the world to which the observations refer. Quantum physicist Niels Bohr is said to have advised his colleagues to do away with philosophy altogether: they should put a sign on the door of their laboratory, "work in progress, philosophers keep out."

But the pragmatic strategy has led to vexing paradoxes. In string theory, for example, correlating the observations calls for assuming more than four dimensions in spacetime: ten or eleven dimensions are required by the mathematics. String theorists assume that there were this many dimensions at the origins of the universe, but in the period of inflation following the explosion that created the universe all but four had been "choked off." But it turned out to be difficult, if not impossible, to "compactify" the extra dimensions without collapsing the remaining four as well.

The paradoxes faced by string theory are not limited to the problem of dimensions: they also concern the reality of the strings themselves. Except in Bohr's Copenhagen school of quantum mechanics, scientists assume that their theories refer to a world that exists independently of their theories. This world must be realistically conceivable, even if it is beyond the scope of commonsense conceptions of the nature of reality.

String theory has difficulty in producing a concept that is realistic by any standard. Strings and superstrings are vibrations, similar to musical notes. However, musical notes are produced by vibrating strings, pipes, or sounding boards, and these are physically real. String theory's strings, by contrast, float in geometrical spacetime in an unsubstantial manner reminiscent of the grin of the Cheshire cat. We have the grin of the cat (the vibration), while its body (the medium that would vibrate and create the various vibrational modes) is not grasped by the theory.

String theory tells us that the vibrational modalities are generated by the geometrical fabric of spacetime, since it is spacetime that undergoes pinching and tearing as Calabi-Yau hole intersections define black holes, wormholes, and elementary particles. Yet spacetime is a geometrical construct, and it is not conceivable that such a construct would vibrate, and by its vibration produce physical effects. As Roger Penrose pointed out, the reality problem of string theory is both that it lacks background geometry and that it calls for multiple untestable dimensions (Penrose 2004).

An analogous paradox emerges in regard to the spacetime concept of relativity theory. In the special theory, light is conceived as the propagation of streams of photons (or alternatively, as the deformation of the spacetime matrix), but the question, "propagation in or through what," or "deformation of what," leads to a paradox. In a humanly conceivably realistic context we would assume that if we have streams or waves traveling from one point in space to another, there is something in nature that extends between these points and carries the streaming entities or waves. The general theory of relativity comes close to answering this query, for it postulates a four-dimensional matrix that carries signals across space. In that matrix gravitation is the principal factor; it is dynamically analogous to Newton's concept of space. However, general relativity's gravitational field is "background independent." The phenomena it describes or postulates are not positioned within an enduring and realistically comprehensible background. Given Einstein's insistence that his theory refers to an independently existing universe, this is paradoxical.

Stephen Hawking's shift from the basic realism that affirms that the universe exists independent of our theories to a "model-dependent realism" that denies an observer-independent theory and a theory-independent reality reflects the prevailing attitude in the physics community (Hawking and Mlodinow 2010). Realism in contemporary physics is further impeded by the persistent gap between the quantum description of the supersmall scale of reality and the universe's relativistic postulates. The force of gravity is not satisfactorily quantized, and quantum gravity is not unified with quantum field theory. It is not clear whether the universe constitutes a smooth spacetime continuum, or it is intrinsically quantized.

The reality of the fields postulated in science is at best ambiguous. But the question regarding the reality of fields goes beyond the status of the fields themselves: the entire reality-concept of contemporary science is at stake. If fields, and the other entities postulated in science, are to be viewed as part of the real universe, we need a conception of reality that is consistent with the observed facts and offers the simplest consistent explanation of these facts. Such a conception has been present in the history of thought, and we may need to revive it in the context of contemporary science. The required conception is that of a dimension that underlies the observed facts.

THE CONCEPT OF A DEEP DIMENSION

Fields, we have noted, are not observables: only their effects can be observed and measured. They share this quality with all the laws and regularities of nature. We observe a dynamically evolving, actualizing universe, but we do not observe the laws and regularities that "drive" it. Cause and effect cannot be collapsed because the effect is manifest while the cause is not—or it is only indirectly so.

The helpful metaphor to elucidate this state of affairs refers to electronic information-processing systems. The hardware of these systems is observable, but—at least in normal operation—their software is not. The software is a set of algorithms programmed into the hardware; it is what makes the hardware behave the way it does. In everyday use we can only deduce the nature, and even the existence, of the software by observing the behavior of the hardware.

The relationship between the software of the system and its hardware holds in regard to the reality of the fields postulated in science. We observe "real world" entities—quanta and quanta-based systems—and note that they are interconnected across space, and possibly over time. We do not observe the fields themselves. However, that fields are unseen is not a warrant for refusing to accept that they may be real. It is a warrant, on the other hand, for maintaining that they exist on a plane of reality that is not the same as the plane of observation.

Fields, and the other forces and laws recognized in science, may exist on a plane or dimension of reality that is "hidden" in regard to direct observation. This assumption has important historical antecedents. Scores of philosophers maintained that the observed world is rooted in a real but unobservable dimension. Philosophers of the mystical branch in Greek metaphysics—the Idealists and the Eleatic school (including thinkers such as Pythagoras, Plato, Parmenides, and Plotinus)—differed on many points but were united in the affirmation of a "hidden" dimension. For Pythagoras this was the Kosmos, a trans-physical, unbroken wholeness, the prior ground on which matter and mind and all being in the world arises. For Plato it was the realm of Ideas and Forms, and for Plotinus "the One." As the Lankavatara Sutra in Indian philosophy affirmed, the deep reality is the "causal dimension" that gives rise to the "gross" phenomena that meets the eye. The world we observe is illusory, ephemeral, and short-lived, while the deep dimension is real, eternal, and eternally unchanging.

At the dawn of the modern age Giordano Bruno brought the concept of a deep dimension into the ambit of modern science. The infinite universe, he said, is filled with an unseen substance called *aether* or *spiritus*. The heavenly bodies are not fixed points on the crystal spheres of Aristotelian and Ptolemaic cosmology, but move without resistance through this unseen cosmic substance under their own impetus.

In the nineteenth century Jacques Fresnel revived this idea and called the space-filling but in-itself unobservable medium *ether*. The ether, he said, is a quasi-material substance in which the movement of heavenly bodies produces friction; it is not observable in itself, but the "ether drag" it produces should be observable. Shortly after the turn of the twentieth century Albert Michelson and Edward Morley tested this assumption. They reasoned that given that the Earth moves through the ether, the light that reaches it from the Sun must display an ether drag: in the direction toward the light source the beams should be reaching the Earth faster than in the opposite direction.

However, the experiments carried out by Michelson and Morley failed to detect a drag that would testify to the presence of the ether. The physics community took this as evidence that the ether does not exist—notwithstanding Michelson's warning that the experiments disproved only a particular mechanistic theory of the ether and not the concept of an invisible space-filling medium that would transport light as well as other fields and forces.

When Einstein published his special theory of relativity, the theory of the ether was discarded: it was no longer necessary. All movement in space—more exactly, in the four-dimensional spacetime continuum—was said to be relative to the given reference frame. It was not to be conceived as movement against a fixed background.

However, the ether, as an unobservable plane of reality underlying the observable phenomena, came back to physics through the back door. Theoretical physicists began to trace the fields and forces of nature to common origins in a unified, later in a grand-unified, then in a super-grand-unified field. For example, in the standard model of particle physics, the basic entities of the universe are not independent material things even when they are endowed with mass; they are part of the unified matrix that underlies space. The basic entities of the matrix are quantized: they are elementary or composite quanta. The elementary quanta include fermions (quarks, leptons, and their antiparticles), and gauge bosons (photons, W and Z bosons, and gluons). Since the autumn of 2012 they also include the previously hypothetical but now experimentally confirmed Higgs boson.

Quanta can be described as waves or as corpuscles. In the wave description (often held to be the more fundamental), quanta are patterns in a field that has nonzero strength even when it is apparently empty. In this field particles acquire mass by interaction with Higgs bosons, the smallest possible excitations of the Higgs field. The Higgs field interacts with particle fields proportionately to the energy carried by the latter. The particle fields, as well as the Higgs field, are manifestation of an extended fundamental field: the unified, grand-unified, or super-grand-unified field. By this token an in-itself unobservable field has emerged as the fundamental matrix of the universe.

The discovery of a new state of matter in the fall of 2012, known as the FQH (Fractional Quantum Hall) state, underscores the concept that everything we experience as "matter" is an excitation of an underlying cosmic matrix. According to theories advanced by Ying Ran, Michael Hermele, Patrick Lee, and Xiao-Gang Wen of MIT, the entire universe is made up of excitations that satisfy Maxwell's equations for electromagnetic waves and Dirac's equations for electrons. Their theory recognizes that in liquids the position of electrons is random and in solids it is rigidly structured. In the FQH state, however, the position of electrons is random at any given time, but in an extended period electrons "dance" in an organized manner. Different patterns of the "electron dance" produce different states of matter.

In the model put forward by Xiao-Gang Wen of MIT with Michael Levin of Harvard (Merali 2007), electrons—as other particles—are the ends of strings that move freely in the underlying medium, "like noodles in a soup." They are woven into "string-nets." Electrons are the ends of strings in the net that fills space. The different patterns in the behavior of the strings account for electrons and for EM waves, as well as for the quarks that make up protons and neutrons and the particles, such as gluons and W and Z bosons, that make up the fundamental forces of nature.

According to Wen, the quantum vacuum is a string-net liquid. Particles are entangled excitations—"whirlpools"—in the space-filling string-net liquid. Empty space corresponds to the ground state of the liquid, and excitations above the ground state constitute particles. The universe is a lattice-spin system constituted of the excitations that manifest as photons and electrons and other (matrix embedded and thus no longer "elementary") particles.

Physicists describe the domain that underlies and embeds the particles, fields, and forces of the universe variously as quantum vacuum, physical spacetime, "nuether," zero-point field, grand-unified field, cosmic plenum, or string-net liquid. However, a revolutionary discovery published in September 2013 places in question even these concepts in regard to their adequacy to describe physical interaction in the universe. The new discovery—the geometrical object known as the *amplituhedron*—suggests that the domain we know as spacetime is not the fundamental reality. The amplituhedron, a mathematical representation of these relationships, is not "in" spacetime; it "governs" spacetime—in the sense in which a computer program governs the entities and relations of that program. It appears that spatio-temporal phenomena are the consequence of geometrical relationships in a deeper dimension of physical reality.

The amplituhedron is a welcome development in quantum-field physics, because it offers an enormous simplification of the calculation of the scattering amplitudes in particle interactions. Previously, the number and variety of the particles that result from the collision of two or more particles (the scattering amplitude of that interaction) have been calculated by so-called Feynman diagrams (diagrams originally proposed by Richard Feynman in 1948). But the number of diagrams required for the calculation is so large that even simple interactions could not be fully calculated until powerful computers came online. For example, describing the scattering amplitude in the collision of two gluons—which results in four less-energetic gluons—requires 220 Feynman diagrams with thousands of terms. Unitl the last few years it was considered too complex to be carried out even with the help of supercomputers.

In the mid-2000s another approach surfaced to calculate scattering amplitudes. Patterns emerged in the description of these interactions that indicate the presence of a coherent geometrical structure. This structure was initially described by BCFW (Ruth Britto, Freddy Cachazo, Bo Feng and Edward Witten) recursion relations. The BCFW diagrams abandon variables such as position and time and substitute for them strange variables—called "twistors"—that are beyond space and time.

They suggest that in the nonspacetime domain two fundamental tenets of quantum field physics—and of contemporary physics as a whole—do not hold: *locality* and *unitarity*. Particle interactions are not limited to local positions in space and time, and the probabilities of their outcome do not add up to one.

The discovery of the geometrical object called amplituhedron is an elaboration of the geometry suggested by the BCFW twistor diagrams. The work of Nima Arkani-Hamed of the Institute for Advanced Study and his former student Jaroslav Trnka, this discovery implies that spacetime, if not entirely illusory, is not fundamental: it is the result of geometrical relationships at a deeper level (Arkani-Hamed 2012; Trnka 2013).

Henceforth physicists can calculate the scattering amplitude of particle interactions in reference to a geometrical object of as many dimensions as the interactions it describes. In principle, a multidimensional amplituhedron could enable the computation of the interaction of all quanta in spacetime. And not only of quanta, but of all complex systems constituted of integrated sets of quanta (living organisms, ecologies, solar systems, galaxies). These interactions are seen as obtaining beyond spacetime: the spacetime features, including locality and unitarity, are consequences of the interactions.

A domain beyond spacetime, familiar in the history of science and philosophy, has resurfaced at the cutting edge of science as the unchanging matrix of the entities and events that populate space and time.

5

The Akasha

THE IDEA OF THE AKASHA

The tenet that the observed world is a manifestation of a deeper dimension is now rediscovered at the cutting edge of quantum field physics. It is not new: it has been a basic element in classical Indian philosophy. Samkhya, one of the earliest philosophical teachings in India, held that there is a compendium of knowledge and information conserved in a nonphysical plane of reality, referred to as the Akashic Records.

The Hindu *rishis* (seers) specified this concept as a full-fledged cosmology. They held that there are not four but five elements of the cosmos. These are *Vata* (air), *Agni* (fire), *Ap* (water), and *Prithivi* (earth)—and also *Akasha*, variously described as space, brilliance, or all-encompassing light. The Akasha is the fundamental element. It holds the other elements in itself, but it is also outside of them, for it is beyond space and time. According to Paramahansa Yogananda, the Akasha is the subtle background against which everything in the material universe becomes perceptible.

In his classic *Raja Yoga*, Swami Vivekananda (1982) gave the following account of the Akasha:

It is the omnipresent, all-penetrating existence. Everything that has form, everything that is the result of combination, is evolved out of this Akasha. It is the Akasha that becomes the air, that becomes the liquids, that becomes the solids; it is the Akasha that becomes the sun, the earth, the moon, the stars, the comets; it is the Akasha that becomes the human body, the animal body, the plants, every form that we see, everything that can be sensed, everything that exists. It cannot be perceived; it is so subtle that it is beyond all ordinary perception; it can only be seen when it has become gross, has taken form. At the beginning of creation there is only this Akasha. At the end of the cycle the solid, the liquids, and the gases all melt into the Akasha again, and the next creation similarly proceeds out of this Akasha (*Raja Yoga*, p. 33).

The Akasha is not merely one element among others; it is the fundamental element: the ultimately real dimension of the cosmos. It is what in its subtle aspect underlies all things and in its gross aspect becomes all things. In its subtle aspect it cannot be perceived. But it can be observed in its gross aspect, in which it has become the things that arise and evolve in space and time. The same concept is present in the Upanishads. "All beings arise from space, and into space they return: space is indeed their beginning, and space is their final end." (Chandogya Upanishad I.9.1)

David Bohm (Nichol 2003) enunciated an identical concept:

What we experience through the senses as empty space is the ground for the existence of everything, including ourselves. The things that appear to our senses are derivative forms and their true meaning can be seen only when we consider the plenum, in which they are generated and sustained, and into which they must ultimately vanish.

In contemporary science space is rediscovered as the fundamental matrix in which the manifest things and events of the universe arise, in and through which they evolve, and into which they again redescend.

THE HOLOGRAPHIC SPACETIME

As we have seen, the deep dimension has been recognized in a number of traditional cosmologies, perhaps the most remarkably in the Akasha concept of the ancient rishis. In Hindu philosophy the world we experience is not the ultimate reality: it is only a manifestation of that reality.

In Akashic cosmology we add a further element to this perennial insight. The deep dimension, we maintain, is beyond spacetime. It creates the holographic spacetime in which we live and which we observe. This conception has now received significant experimental support.

Evidence for the holographic nature of spacetime has surfaced in the spring of 2013. As reported in *New Scientist,* Fermilab physicist Craig Hogan suggested that the fluctuations observed by the British-German gravity-wave detector GEO600 may be due to the graininess of spacetime (as noted, according to string theory, at the supersmall scale spacetime is patterned by minuscule ripples: it is "grainy"). The GEO600 gravity wave detector did find inhomogeneities in the matrix that constitutes spacetime, but they were not gravity waves. Could it be, Hogan asked, that they are the ripples that string theory claims pattern the microstructure of spacetime? This would be the case if the micro-inhomogeneities are 3D projections of information coded in 2D at the circumference of spacetime. In that event we could assume that events within spacetime are 3D projections of 2D information encoded at the periphery.

The holographic spacetime hypothesis has been revived to account for the anomaly connected with the "evaporation" of black holes. In the 1970s Hawking (1974) found that, as black holes evaporate, the information contained in them is lost. All the information about the star that had collapsed and become a black hole vanishes. This is a problem, because information, according to contemporary physics, cannot be lost in the universe.

This problem was resolved when Hebrew University cosmologist Jacob Bekenstein discovered that the information present in the black hole (a quantity equal to its entropy) is proportional to the surface area of its event horizon, the horizon beyond which matter and energy cannot escape. Physicists have shown that quantum waves at the event horizon encode the information present in the black hole. This information is proportional to the volume of the black hole; thus there is no unaccounted-for information loss as the black hole "evaporates."

Leonard Susskind and Gerard 't Hooft applied the principle of information coding to spacetime as a whole. They pointed out that spacetime has an event horizon of its own: it is the circumference of the area that light has reached in the period since the birth of the universe. Juan Maldacena demonstrated that the physical properties of a 5D universe are identical with the coding of its 4D spacetime boundary. The coding is in "bits": each Planck dimensional square at the boundary codes one bit of information. This theory resolves the problem of black hole information loss in the universe, but it is not observationally verifiable: events at the Planck scale are too small to be observed.

Applying the holographic coding theory to the whole of spacetime overcomes the verifiability problem. Since the volume of the universe is larger than its surface (we can calculate the difference by dividing the surface area by the volume), then it follows that if the 2D codes that project 3D events within spacetime occupy a Planck dimensional area on the surface, the three-dimensional events they code must be of the order of 10^{-16} and not of 10^{-35} centimeter. Events of this larger dimension are observable. If the ripples found by the GEO600 gravity wave detector are of the order of 10^{-16} cm, they could be ripples in the microstructure of spacetime. Observations made at the time of writing indicate that this is indeed the case.

Further support for holographic spacetime theory came in the fall of 2013 when Yoshifumi Hyakutake and colleagues at Ibaraki University in Japan computed the internal energy of a black hole, the position of its event horizon, its entropy, and several other properties based on the predictions of string theory and the effects of virtual particles. Hyakutake, together with Masanori Hanada, Goro Ishiki, and Jun Nishimura, also calculated the internal energy of the corresponding lower-dimensional cosmos with no gravity. They found that the two calculations match (Masanori Hanada, Yoshifumi Hyakutake, Goro Ishiki, Jun Nishimura, 2013). It appears that black holes, as well as the cosmos as a whole, are holographic. The microstructure of space is patterned by 3D ripples that correspond to 2D codes at the space-time periphery, and the internal energy of a black hole and the internal energy of the corresponding lower-dimensional cosmos are equivalent. This suggests that spacetime is a cosmic hologram, and that quanta and systems constituted of quanta are intrinsically entangled elements of it.

The dimension that generates the holographic spacetime we experience is the Akasha. The Akasha harbors the geometrical relations that govern the interaction of quanta and of all things constituted of quanta in space and time. It is the seat of the fields and forces of the manifest world. The Akasha is the universal gravitational field that attracts things proportionately to their mass; it is the electromagnetic field that conveys electric and magnetic effects through space; it is the ensemble of the quantum fields that assigns probabilities to the behavior of quanta; and it is the scalar holofield that creates nonlocal interaction among quanta and configurations of quanta. The Akasha is the integration of all these elements in a unitary cosmic dimension that is beyond space and time. It is the fundamental, if in the everyday context hidden, dimension of the world.

PART TWO

Akasha Paradigm Cosmology

Having outlined the conceptual foundations of the new paradigm in science, we now discuss the implications of that paradigm for our understanding of the world—first in its largest dimensions. We include in this discusion the phenomena of consciousness. In a conception based on the Akasha paradigm consciousness is an integral part of the coherent totality the Hellenic philosophers called *cosmos*.

6

Cosmos

In natural science cosmology is an empirical inquiry, searching for understanding the origins, evolution, and destiny of the macro-structures of the universe grounded in observation and experiment. In the context of philosophy cosmology is a broader inquiry: it intersects with metaphysics (the science of first principles, based on the fundamental "physics" of the world) and with ontology (systematic inquiry into the nature of reality). Here we take cosmology in the broad philosophical context, but with due regard for the findings that come to light in natural science.

NEW HORIZONS IN COSMOLOGY

Conceptions of the nature of the cosmos have changed throughout history. Cosmological theorizing has been highly sensitive to the chosen paradigm. When that paradigm was mechanistic, the resulting cosmology portrayed the world as a giant mechanism. When the paradigm was vitalistic, the picture that emerged was of the world as a cosmic organism. And when the paradigm was idealistic, the reality it perceived appeared as the manifestation of a cosmic mind or consciousness.

For the past three hundred and fifty years, Western science has been dominated by the materialistic Newtonian paradigm. Cosmologies based on this paradigm envisaged the universe as a vast mechanism, running on the energy—the negative entropy—with which it was endowed at its birth. The universe was believed to be inexorably running down, toward the state of maximum entropy where irreversible processes are no longer possible and no life can come about, indeed nothing other than the repetition of reversible processes.

However, the staggering energy sea discovered at the quantum level of the universe challenged the Newtonian concept of a closed clockwork universe. Another concept has been emerging, grounded in the insight that a deep, quasi-infinite medium or matrix subtends the world we observe. In new-paradigm cosmology we identify this matrix as the Akasha. Here we present the basic tenets of the new cosmology in the form and the language appropriate to discourse about the fundamental nature of reality.

FIRST PRINCIPLES

The cosmos is an integral system actualizing in the interaction of two dimensions: an unobservable deep dimension and an observable manifest dimension. The deep dimension is the Akasha: the "A-dimension." The observable dimension is the manifest "M-dimension." The A- and the M-dimensions interact. Events in the M-dimension structure the A-dimension: they alter its potential to act on—to "in-form"—the M-dimension. The A-dimension "in-forms" the M-dimension, and the in-formed M-dimension acts on—"de-forms"—the A-dimension. The M- and the A-dimensions do not signify a cosmos split in two. The cosmos is one, but for the observer it is meaningfully considered under the heading of two dimensions: a fundamental dimension and an experienced dimension. The diversity of events in the experienced dimension is a manifestation of the unity that governs their interaction in the fundamental dimension. This is the basic tenet of Akashic cosmology. We now elaborate it in more detail.

The particles and systems of particles that arise in the M-dimension interact with each other as well as with the A-dimension. Every particle and every system of particles has what philosopher Alfred North Whitehead called a "physical pole" through which it is affected by other particles and other systems of particles in the M-dimension, and a "mental pole" by which it is affected by the A-dimension. Whitehead called these "prehensions"—the action of the rest of the world on particles and systems of particles in space and time.

As all things in the M-dimension, human beings have both a physical pole and a mental pole. We "prehend" the world in two modes. We prehend the M-dimension through the fields and forces that govern existence in the manifest world, and we prehend the A-dimension as the spontaneous intuitions Plato ascribed to the realm of Forms and Ideas, Whitehead to eternal objects, and Bohm to the implicate order. The former are the known effects of the external world on our organism, and the latter the more subtle insights and intuitions that appear for most of us but are mostly ignored in the modern world.

The M-dimension and the A-dimensions are related diachronically (over time) as well as synchronically (at a given point in time). Diachronically, the A-dimension is prior: it is the generative ground of the particles and systems of particles that emerge in the M-dimension. Synchronically, the generated particles and systems of particles are linked with the A-dimension through bidirectional interactions. In one direction the A-dimension "in-forms" the particles and systems that populate the M-dimension; and in the other the in-formed particles and systems "de-form" the A-dimension. The latter is not the eternal and immutable domain of Platonic Forms, but a dynamic matrix progressively structured by interaction with the M-dimension.

Particles and systems of particles in the M-dimension are not discrete entities, disjoined either from each other or from the A-dimension. In the last count the things that populate the M-dimension are soliton-like waves, nodes, or crystallizations of the A-dimension. It is in the A-dimension that they actualize, it is with the A-dimension that they "dance" and coevolve. And it is into the A-dimension that they redescend when a universe completes its evolutionary/devolutionary trajectory in the multiverse.

THE METAPHOR OF SEA AND WAVES

We can employ a simple metaphor to illustrate the reciprocal relation of the M- and the A-dimensions. We refer to the sea. If there is no wind or other disturbance, its surface is still and smooth. But as soon as something perturbs the surface, waves appear on it. These waves are not separate realities: they are part of the body of water—they are its surface manifestations. If we concentrate on the waves, we see wavetrains propagating, interacting and creating complex patterns. Yet the waves are patterns produced by the body of water that makes up the sea. The waves are not *in* or *on* the sea: they are waves *of* the sea. They are the manifestation of the reality of the body of water.

We now add a further element to this metaphor: the element of interaction. The body of water forms the waves that appear on its surface, and the waves deform in turn the body of water. Here the body of water is the A-dimension, and the waves that appear on its surface are events in the M-dimension.

THE ORIGINS OF COHERENCE IN THE UNIVERSE

Recent cosmological models view the universe as a cycle in a vaster and possibly infinite "multiverse." The universe we inhabit is not "the" universe but merely a "local" universe.

Being a cycle in a vaster multiverse offers a cogent explanation of the coherence that characterizes our own universe. Our universe is astonishingly coherent: all its laws and parameters are finely tuned to the emergence of complexity. If the universe were any the less coherent, life would not be possible and we would not be here to ask how life had evolved on Earth and possibly elsewhere in the vast reaches of cosmic space. Did it evolve by chance? Or by design? The most plausible answer is that the coherence of our universe is not due either to serendipity or to supernatural design. It is due to trans-universe *inheritance*.

The hypothesis of serendipity faces a serious problem of probability. Although the theory of large numbers allows that in a large number of tries even otherwise improbable outcomes have a reasonable probability of coming about, the number of tries needed to reach a significant probability that a coherent universe such as ours would come about is enormously high. The "search space" for this selection is the number of universes that are physically possible, and according to some versions of string theory, this number is of the order of 10^{500}. On the other hand the number of "hits" is extremely limited: only a handful of this staggeringly large number of possible universes is capable of bringing forth life; the rest are biologically sterile. And yet life has evolved on this planet, and it may exist in this universe's other planets as well.

The hypothesis of design appeals to supernatural providence and invites the question, design by what? Or by whom? If the answer is that the Designer is not part of the natural universe, the argument shifts to the domain of theology. It may be a good answer, perhaps even "the" answer, but discussing it is beyond the ken of science.

On the other hand, the hypothesis of inheritance—more exactly, that our universe would have inherited its coherence-generating properties from a precursor universe—is now well within the domain of science. Sophisticated cosmological models lend a measure of theoretical support for it. One such model is the loop quantum gravity cosmology developed by Abhay Ashtekar and a team of astrophysicists at the Institute for Gravitational Physics and Geometry at Penn State University (Ashtekar et al., ed. 2003). Loop quantum gravity cosmology permits the definition of the state of our universe prior to the "Bang" that gave rise to it.

The standard model does not permit going back even to the time that immediately followed the Big Bang: matter was so dense then that the equations of general relativity did not hold. The mathematics of loop quantum gravity now permit "retrodicting" the conditions that reigned in the universe not just immediately after its explosive birthing but also prior to it. In this model the fabric of space is a weave composed of one-dimensional quantum threads. In this model Einstein's four-dimensional continuum is only an approximation; the geometry of spacetime is not continuous but has a discrete "atomic" structure. Before and during the primal explosion this fabric was torn apart, making the granular structure of space dominant. Gravity shifted from a force of attraction to a force of repulsion, and in the resulting cosmic explosion our universe was born.

The simulations of loop quantum gravity indicate that prior to the birth of our universe there was another universe with physical characteristics similar to our universe. Ashtekar and collaborators were surprised at this finding and kept repeating the simulations with different parameter values. But the finding held up. It appears that our universe was not born in the singularity known as the Big Bang, and it will not end in the singularity of a Big Crunch. The universe we inhabit is not "our" universe but a "multiverse" of an indefinite number of universes. Big Bangs and Big Crunches are phase transitions in the multiverse, critical transitions from one universe to the next, with spacetime shrinking to quantum dimensions. The matter-component of the prior universe "evaporates" in black holes and is reborn in the superfast expansion that follows the final collapse. Instead of a Big Bang leading to a Big Crunch, we have recurring Big Bounces.

The theory of inheritance between successive universes (successive cycles of the multiverse) offers a more cogent explanation of the coherence of our universe than either the theory of random selection or a theological appeal to supernatural providence. Calculations based on loop quantum gravity cosmology lend the inheritance thesis significant support. It appears that in the transition between the successive cycles the parameters and other physical characteristics of the preceding cycle are not canceled but affect the next cycle.

Calculations by Alejandro Corichi of the National University of Mexico (Corichi 2007) show that a "semiclassical state" on one side of the Big Bounce peaks on a pair of canonically conjugate variables and these strongly bound fluctuations on the other side. The change in fluctuations on the two sides is insignificantly small (10^{-56}) even for a universe of 1 megaparsec and becomes still smaller for larger universes.

The hypothesis of trans-cyclic inheritance is compatible with some versions of superstring theory. An evolved form of M-theory calls for eleven dimensions: ten dimensions of space, and one of time. According to Brian Greene our universe is a "three-brane set" embedded within a larger string-landscape consisting of many three-brane sets (*branes* are mathematically defined spatially extended entities that may have any number of dimensions). The branes collide and create a rebound. This dynamic drives the evolution of universe cycles in the multiverse.

As discussed in appendix I, a number of cosmological models embrace the concept of an enduring cosmic matrix underlying the birth, evolution, and devolution of our universe and other local universes. The hypothesis of trans-cyclic inheritance suggests that these universes are not born in a condition of tabula rasa: through the underlying cosmic matrix they are "in-formed" by the collapsing precursor universe.

EVOLUTION IN THE UNIVERSE

Astronomical and astrophysical evidence leaves no doubt that the manifest dimension of the cosmos—the spacetime of the universes that emerge in the multiverse—evolves over time. The matter-content of these universes is created in Big Bangs and vanishes in Big Crunches. In between it evolves in stars and galaxies, and reaches high levels of complexity on physically favorable planetary surfaces.

Akashic cosmology adds further specifics to this standard model. The particles that arise in each cycle of the multiverse evolve to form systems of particles, but they cannot evolve indefinitely. Evolution in each cycle is limited by physical conditions in that cycle. The universe-cycles are finite, and the conditions they provide are not indefinitely conducive to the persistence of complex systems.

Thermal and chemical conditions in each universe are suitable for the buildup of complex systems only during the expansionary phase. As expansion reaches an apex and gives way to contraction, physical conditions for complex systems become unfavorable. During the supercompacted phase in a collapsing universe only the stripped nuclei of atoms persist; then they also die back into the quantum vacuum, the cosmic plenum.

However, explosive birthing followed by expansion followed in turn by contraction and collapse does not describe the full range of the evolution that takes place in the cosmos. The cosmos is a multiverse of successive universes, and the evolution of one universe is the evolution of only one cycle in the multiverse. On the theory of trans-universe inheritance, the physical characteristics of one universe affect the physical characteristics of the next.

In this fashion a vaster process of evolution unfolds in the cosmos. In each cycle the A-dimension of a universe in-forms the M-dimension, and the in-formed M-dimension deforms the A-dimension. Thereby a learning curve obtains across the cycles. The A-dimension is progressively de-formed, and it progressively in-forms the M-dimension. Consequently the systems that populate the M-dimension are increasingly in-formed by the A-dimension. Thus they attain higher peaks of evolution in equal times, or equal peaks of evolution in shorter times.

Where does this process lead? If the sequence of cycles produces progressive evolution in the M-dimension, and if that evolution deforms the A-dimension, the sequence of cycles must ultimately reach a zenith—a final omega cycle. In the omega cycle the A-dimension actualizes its full potentials to in-form systems in the M-dimension, and systems in the M-dimension achieve the highest level of evolution that is physically possible in space and time.

PARALLELS WITH HINDU COSMOLOGY

The interaction outlined in Akasha paradigm cosmology between an enduring fundamental reality (the A-dimension of the multiverse) and a transient phenomenal world (the manifest universes in the multiverse) has been anticipated in Hindu cosmology. For classical Hindu thought the almost infinitely varied things and forms of the manifest world are reflections of oneness at the deeper nonmanifest level. At that level the forms of existing things dissolve into formlessness, living organisms exist in a state of pure potentiality, and dynamic functions condense into static stillness. All attributes of the manifest world merge into a state beyond attributes. Time, space, and causality are transcended in Brahman, the state of pure being.

Brahman, though undifferentiated, is dynamic and creative. From its ultimate "being" comes the temporary "becoming" of things in the manifest world, with their various attributes, functions, and relationships. The *samsara* of being-to-becoming, and then of becoming-to-being, is the *lila* of Brahman: its play of creation and dissolution. The absolute reality of Brahman and the derived reality of the manifest world form an interconnected whole: together they are the *advaitavada* of the cosmos. Brahman is the fundamental reality. The things that appear in the manifest world have secondary reality—mistaking them for real is the illusion of *maya*.

In Akashic cosmology the A-dimension takes the place of Brahman: it is the ultimate reality of the cosmos. The universes that evolve and devolve in space and time are manifestations of the Akasha, the enduring matrix of the multiverse. They are the breathing in and the breathing out of Brahman—the inflation and the deflation of the universe created in and by the Akasha.

7

Consciousness

Is consciousness a universal phenomenon, a part of the reality of the cosmos? Or is it limited to living species, and possibly to the human species alone? All these alternatives have been affirmed in the history of philosophy. Choosing between them, the same as in regard to other questions about the nature of reality, needs to be based on the consistency and coherence of the scheme by which the observed facts are tied together.

The new paradigm offers a simple, consistent, and coherent scheme that can tie together the facts regarding the "material" reality of the cosmos and the "immaterial" phenomena of mind and consciousness.

THE BODY-MIND PROBLEM

Consciousness is present in association with the human brain, whether or not it is also present in the wide reaches of the cosmos. The classical problem is how this consciousness is related to the human brain, and through the brain to the rest of the world.

Brain and mind are both elements of human experience. We experience things that appear material around us, and we experience the apparently immaterial phenomenon we call mind or consciousness. These experiences cannot be entirely independent; they are both part of the flow of human experience. But how is the experience of "matter" related to the experience of "mind"?

Although the experience of matter and the experience of mind are both part of the flow of human experience, they are fundamentally different parts. They differ in the way they appear and differ also in the way they can be accessed. Matter, it seems, can be experienced by all people, and perhaps all systems endowed with some form of sensitivity to their surroundings. On the other hand, consciousness is an intensely private experience, available only to the experiencing subject. As skeptical philosophers pointed out, the existence of consciousness in other people is a conjecture based on our own experience of consciousness. But *does* matter have an independent reality, and is consciousness associated also with things *other* than the human brain? These questions have been debated for millennia, and while no definitive answer has emerged, the principal positions that could provide an answer have crystallized. We state them here under the headings of materialism, idealism, and dualism.

> For *materialism* all the things that exist in space and time are material—they are made up of a substance called matter.
>
> For *idealism* all things in the world are mental, or at least mind-like. Mind and consciousness are the basic and possibly the sole reality.
>
> For *dualism* both matter and mind are real. Humans, and perhaps all living things, are material as well as mental.

All these positions claim empirical support.

Materialists point out that the world to which our experience refers is a world of solid things, such as atoms and molecules, and the many things composed of atoms and molecules. These are material things: they are "matter" in one combination or another. Mind and consciousness are epiphenomena, by-products of the brain, which is a complex material thing in its own right.

Idealists claim that we experience the world through our consciousness, and all we know of the world comes to us through our consciousness. Our consciousness consists of a stream of perceptions, volitions, feelings, and intuitions, and although some of these things seem to refer to things outside our consciousness, in the final count we cannot be sure that they are not just items of our consciousness. The philosopher Descartes analyzed the stream of human consciousness and failed to find convincing proof that there is a world that exists independently of it. The only thing he could not doubt is the act of experiencing itself—*cogito ergo sum*: I think, therefore I am.

Dualists believe that both matter and mind are fundamental elements of the world. Whether or not mind is present in all things or only in humans, in humans it is associated with a brain and nervous system.

Despite volumes having been written in defense of one or another of these alternatives, the relation between a material brain and an immaterial consciousness remained an unsolved problem. Philosopher David Chalmers (1995) called this the "hard problem" of consciousness research.

> When we see, for example, we experience visual sensations: the felt quality of redness, the experience of dark and light, the quality of depth in a visual field. Other experiences go along with perception in different modalities: the sound of a clarinet, the smell of mothballs. Then there are bodily sensations, from pains to orgasms; mental images that are conjured up internally; the felt quality of emotion; and the experience of a stream of conscious thought. What unites all of these states is that they all are states of experience. (Chalmers 1995)

Being in a state where we experience the qualitative flow of sensations is fundamentally different from observing the states of a material entity such as the brain. Does that material entity produce the immaterial flow we know as our consciousness? But how could a material entity produce something immaterial?

This "hard problem" contrasts with the "easy problems" of consciousness research. For example, our ability to discriminate, categorize, and react to environmental stimuli is a relatively easy problem because it can be solved in principle in reference to neural and artificial computational mechanisms. When our brain engages in visual and auditory information processing, we have visual and auditory experiences. The same holds for our understanding of the way the nervous system accesses our body's own states.

But the hard problem persists. How can a network of neurons that process nerve signals from the senses produce felt qualitative experience? Philosopher Jerry Fodor (1992) wrote that "nobody has the slightest idea of how anything material could be conscious, nor does anybody even know what it would be like to have the slightest idea about how anything could be conscious."

Finding a solution to the body-mind problem is indeed hard—at least when the problem is posed in light of the old paradigm. The Akasha paradigm offers a radically different framework for this problem. It suggests that brain and mind do not exist on the same plane of reality, in the same dimension of the cosmos. The brain is a part of the material plane of reality: the manifest M-dimension. Mind and consciousness, on the other hand, participate in, and essentially belong to, the deep Akasha dimension.

A NEW VIEW OF THE PLACE OF CONSCIOUSNESS IN THE COSMOS

We have already noted that the idea of a real but hidden dimension is a major tenet in traditional cosmologies. We have also seen that the same insight has surfaced in the latest theories of science. The existence of this deep dimension is the key to finding the place of consciousness in the cosmos.

The idea that consciousness belongs to another, deeper dimension of reality where all individual consciousnesses are one has been frequently voiced, and not only by poets and prophets. Physicist Erwin Schrödinger said that "the overall number of minds is just one. . . . In truth, there is only one mind" (Schrödinger 1969). Consciousness, he added, does not exist in the plural. In his later years psychologist Carl Jung came to a similar conclusion. He noted that the psyche is not a product of the brain and is not located within the skull; it is part of the generative, creative principle of the cosmos—of the *unus mundus*.

Akashic cosmology is fully aligned with these concepts. It claims that consciousness is not produced by the brain, and it is not part of the physical reality of the manifest world. Consciousness originates in the A-dimension, and it infuses the manifest world by interaction with that dimension. The neural networks of the human brain resonate with the information present in the A-dimension. In a more technical vein we can say that the brain performs the equivalent of a Gabor transform in regard to signals from the A-dimension: it translates the information carried in that dimension in a holographically distributed form into linear signals that affect the functioning of the brain's neural networks. This—nonlocal—information first reaches the subneural networks of the right hemisphere, and then, if it penetrates to the level of consciousness, also reaches the neuroaxonal networks of the left hemisphere. Given that this information is a translation of holographically distributed information in the A-dimension, it conveys the totality of the information in that dimension. Thus our brain is imbued with the totality of the information that pervades the cosmos.

This claim is theoretically sound, but it is not corroborated by experience. Clearly, our consciousness does not display all the information that exists in the world. But this does not mean that such information would not be available to our brain; it only means that our brain filters out all but a tiny segment of this information. In the everyday context we perceive only those aspects of the world that are relevant to our life and aspirations.

The censorship of the brain does not mean an absolute limitation; in nonordinary states of consciousness this limitation can be vastly expanded. The experience of transpersonal psychiatrists and psycho-therapists shows that in nonordinary, altered states of consciousness we can receive information from almost any part of the world, and from almost any time. It appears that, at least in potential, we do have access to the complete and permanent record of all things in space and time—that we can "read" the entire "Akashic Record."

Expanded access to Akashic information accounts for a number of otherwise puzzling phenomena. It explains the seemingly complete long-term memory that comes to light in altered states, including those that accompany near-death experiences. In these instances our brain decodes information from our own past, bringing to consciousness a part of the record of our own past interaction with the world around us. It also explains transpersonal experiences. Given that A-dimensional information is holographically entangled nonlocal information, we should also be able to "read" some elements of other people's consciousness. The findings of transpersonal psychologists, mediums, and particularly gifted and sensitive individuals testify that this claim is not exaggerated. It appears that we can, and sometimes do, read other people's conscious-ness, whether they are alive today or lived sometime in the past.

Here the timeless insights of spirituality converge with the latest findings of consciousness research. The emerging insights are at the same time spiritual and scientific. They can be briefly encapsulated.

I, a conscious human being, am not limited to my body. I am a matter-like system in the manifest dimension of the world, and a mind-like system in the Akasha dimension. As a matter-like system I am my body, and I am ephemeral. But as a mind-like system I am my consciousness, and I am part of the world's deep dimension. I am omnipresent and immortal, a nonlocal part of the infinite wholeness of the cosmos.

PART THREE

Akasha Paradigm Philosophy

In this part we shift from the largest view of the world to the more modest yet still embracing view of humankind as a conscious species arising in and interacting with the rest of the world, and most immediately with other humans in the web of life on this planet.

Akasha paradigm philosophy sheds fresh light on such humanly relevant perennial questions as the nature and scope of our perception of the world, the origins of the vitality inherent in our body, the range and limits of human freedom, and the objectivity and meaning of the human aspiration to achieve the highest value philosophers called "the Good."

8

Perception

The Akasha paradigm recovers an ancient insight: the presence of a deep dimension in the cosmos. This A-dimension is the record and the memory of all the things we experience; it connects all things with all other things; it conserves the trace of all that has already happened and it "in-forms" all that is yet to happen.

In the context of human experience the Akashic deep dimension is a source of intuitions, hunches, creative ideas, and sudden insights. These elements of our experience are not given credence in the modern world; we usually ignore or repress them. Doing so is based on a mistaken understanding of the nature of the world, and of the potentials of our perception of the world.

TWO MODES OF PERCEIVING THE WORLD

The realization that is currently emerging at the leading edge of brain and consciousness research is that we have two sources of information reaching us from the world and not just one. We receive information from the manifest M-dimension, as well as from the deep A-dimension. The information we receive from the M-dimension is in the form of wave propagations in the electromagnetic spectrum and in the air. And the information we receive from the A-dimension is in the form of wave propagations on the quantum level. Signals from the M-dimension are received through our senses, and those from the A-dimension are processed by the quantum-level decoding networks of our brain without passing through the senses.

Everyday experience is dominated by information conveyed by the five senses: these are the sights, sounds, smells, flavors, and textures of the world around us. Until recently most people, including scientists, believed that this is the only information we can obtain from the world. This reduced the scope of our experience to the elaboration of sensory data. New developments in cutting-edge neuroscience show that the classical concept is too narrow; it ignores an essential element of human experience.

Sensory information is processed by neurons connected by synapses in the brain's neuroaxonal network. This network is only one of the systems that process information from the world: there is a vast hierarchy of networks below this level, extending all the way to quantum dimensions. The brain's subneuronal networks are built of cytoskeletal proteins organized into microtubules. They are connected to each other structurally by protein links and functionally by gap junctions. Operating in the nanometer range, the number of elements in these subneuronal networks substantially exceeds the number of elements in the neuroaxonal network: there are approximately 10^{18} subneuronal microtubules in the brain, compared with "merely" 10^{11} neurons.

Microtubules, cylindrical polymers of the protein "tubulin," are major components of the cell's cytoskeleton. They self-assemble into intracellular structures, creating and regulating synapses and communicating between membrane structures and genes in the cell's nucleus. They continually differentiate and re-shape themselves, acting as the cell's nervous system.

Neurophysiologist Stuart Hameroff and physicist Roger Penrose proposed a sophisticated theory of information processing by the brain's microtubular networks (Hameroff 1987; Penrose 1996, Hameroff, Penrose et al. 2011). Microtubule-level information processing raises the brain's information-processing capacity. Rather than a few synaptic bits per second, 10^8 tubulins per neuron switching coherently in the 10^6 Hertz domain produces potentially 10^{14} bits per second per neuron.

Quantum-level processes extend the brain's information-processing capacity to the basic level in the universe. Beyond the level of atoms we have the Planck-level with geometry at 10^{-33} centimeter. This is the level of the fine-structure of space, with graininess, fluctuation, and information. The GEO600 gravity wave detector recorded a fractal kind of noise emanating from fluctuations at this scale. The fluctuations repeat every few orders of size and frequency, from the Planck scale of 10^{-33} per centimeter and 10^{-43} per second, to biomolecular size and time: 10^{-8} cm and 10^{-2} sec. At higher frequencies, such as at 10 kHz, megahertz, gigahertz and terahertz domains, cerebral information processing involves more and more of the brain's microtubules and subneuronal assemblies and may ultimately involve the entire brain (Hameroff and Chopra 2013).

Psychiatrist and brain researcher Ede Frecska and social psychologist Eduardo Luna affirmed that we have two distinct systems in the brain that process information: the classical neuroaxonal network and the quantum-level microtubular network (Frecska and Luna, 2006). The neuroaxonal network gives us the "perceptual-cognitive-symbolic" mode of perceiving the world, and the microtubular network offers a "direct-intuitive-nonlocal" mode. The perceptual-cognitive-symbolic mode dominates consciousness in the modern world; information processed in the direct-intuitive-nonlocal mode is mostly filtered out.

Perception, it appears, is selective in both the classical and in the quantum-processing mode. The brain's collection of nerve cells function as multilayered frequency receptors, selecting the signals to which they respond. Due to conditioning from early in life, each receptor becomes wired to respond to a particular frequency. The act of "tuning in" to the information reaching our brain means picking out the frequency patterns that are familiar from an ocean of patterns and frequencies that are unfamiliar and thus ignored.

As the receptors tune in to particular frequencies, a pattern-recognition response is generated. The information-processing networks interpret the selected pattern in accordance with the interpretation already established for it. By tuning in to the same pattern over and over again, the established interpretation is reinforced.

Selectivity based on repeated patterns is typical for all aspects of our experience: we have difficulty in recognizing, or even perceiving, unfamiliar patterns. This kind of selectivity operates also in regard to the quantum-level signals processed by the brain's subneuronal networks. For most people in the modern world the information received in this mode is unfamiliar, esoteric, and vaguely threatening, and is selectively filtered out.

READING THE AKASHIC FIELD

We can take effective steps to reduce the selective censorship of waking consciousness: we can enter nonordinary states of consciousness. Some nonordinary states obtain spontaneously: they are the hypnagogic and hypnopompic states that mark the transition between sleep and wakefulness. Others can be purposively induced, for example, through rhythmic movements, rhythmically repeated sounds or images, and the controlled use of hallucinogenic drugs. These techniques have been familiar to people in traditional cultures. They are known to and employed by modern psychiatrists and psychotherapists.

The calm contemplation of nature and the enjoyment of poetry, music, and art can bring about altered states of consciousness. A similar end is achieved by religious rituals. Repetitive chanting, drumming, and dancing were the principal ways of inducing altered states in traditional cultures, and their equivalent is the ritual practice of prayer, with religious fervor intensified through the repetition of sacred words, prayers, or invocations.

Yoga is an age-old discipline dedicated to achieving altered states of consciousness. It includes four distinct practices for inducing such states: the familiar forms of meditation, known as the yoga of being; the feeling of love, known as the yoga of feeling; the search for understanding, the yoga of the intellect; and karma yoga, the shift in attitude said to be the yoga of action.

Altered states of consciousness can be catalyzed by traumatic or other life-transforming experiences. Astronauts who have had the privilege of viewing the Earth from outer space often report experiences that go beyond the scope of their bodily senses. This was the case for Edgar Mitchell, captain of the lunar module of Apollo XIV, the sixth man to walk on the moon. While he was in space he had an epiphany that changed his life. He reported that he beheld the network of life on Earth as an interconnected whole. On returning to Earth, he founded the Institute of Noetic Sciences.

Mitchell wrote that the answer to how spontaneous events lead to deep insights with significant behavioral modifications lies in the Akashic field, where it is augmented continuously by quantum holographic information. The intuitions that surface in these experiences are our first rather than our sixth sense, because it was developed long before humans began to ascribe the information they receive to their five bodily senses (Mitchell 1977).

Entering altered states of consciousness, whether through unusual events or through prayer, meditation, esthetic experience, or the contemplation of nature, has a measurable effect on the brain: it synchronizes the left and the right hemispheres. Experiments show that in altered states the electrical activity of the two hemispheres becomes harmonized: the patterns that appear in one are matched by patterns in the other. This contrasts with ordinary states where the two hemispheres function nearly independently of each other.

Brain-wave harmonization is not limited to the brain of the given subject. When several people enter a deeply meditative state together, not only their own hemispheres become synchronized; synchronization extends to the entire group of meditators. In experiments performed by Italian brain researcher Nitamo Montecucco (2000), eleven of a group of twelve individuals in deep meditation achieved a level of transpersonal synchronization that exceeded ninety percent. Yet the meditators sat with closed eyes, in silence, and did not see, hear, or otherwise perceive each other.

Another experiment that testifies to the nonlocal transmission of information from the brain of one person to that of another was carried out in the presence of this writer in southern Germany in the spring of 2001. At a seminar attended by more than a hundred persons, Günter Haffelder, head of the Institute for Communication and Brain Research in Stuttgart, measured the EEG patterns of Dr. Maria Sági, an experienced natural healer, together with that of a test subject who volunteered from among the participants. The subject remained in the seminar hall while the healer was taken to a separate room. Healer and subject were wired with electrodes, and their EEG patterns were displayed on a monitor in the hall. During the time Dr. Sági was diagnosing and treating the subject, her EEG waves dipped into the Delta region between 0 and 3 Hz, with a few eruptions of higher wave amplitude. The subject sat in the hall in a light meditative state without sensory contact with the healer. Yet in his EEG the same Delta wave pattern emerged after about two seconds (Sági 2009).

The reception of information in nonordinary states of consciousness can reach mind-boggling dimensions. According to a number of healers, psychotherapists, and psychiatrists, the experiences of patients in altered states can include spontaneous contact with persons, things, and events they could not have accessed and experienced through their bodily senses. Stanislav Grof found that in altered states people experience a loosening and melting of the boundaries of their ego and a sense of merging with other people and other forms of life. In deeply altered states some people report an expansion of their consciousness to the extent that it encompasses all life on the planet. Individuals perceive sites, persons, and events they say they had experienced in previous lives. So-called psi phenomena occur more frequently in these states, and telepathic and clairvoyant abilities tend to surface.

Grof (2012) is clear about the reality of altered-state perceptions. Having spent more than half a century studying transpersonal experiences, he did not hesitate to assert that he does not doubt that many if not all such experiences are ontologically real and not the product of metaphysical speculation, human imagination, or pathological processes in the brain. Anybody doubting their authenticity, Grof wrote, would have to explain why these experiences have been described by people of various races and cultures and in diverse historical periods, and why they emerge in modern societies under such diverse circumstances as sessions of experiential psychotherapy, meditation, psychospiritual trauma, and near-death experiences.

The classical theory of perception needs to be revised. We receive two kinds of information from the world, and they are processed by two different kinds of systems: information from the M-dimension reaches our bodily senses and is processed by our brain's neuroaxonal network, and information from the A-dimension reaches our brain directly and is decoded in the brain's subneuronal network of networks.

In our critical times we need insights and intuitions of oneness and connection. It is essential to recognize—literally "re-cognize"—that the integral holographic information reaching us from the Akasha is just as real as, and may on occasion be even more valuable than, the sensory information that reaches us from the manifest world.

9

Health

The paradigm emerging in the current revolution in science is of more than scientific interest: it has practical implications. One of these implications—the one that is both the most ancient and the most revolutionary—regards our health. New-paradigm medicine maintains that the health of our body can be sustained, and reestablished when needed, using the information we receive from the Akasha.

INFORMATION IN THE ORGANISM

The human body consists of trillions of cells, and each cell produces thousands of bio-electro-chemical reactions every second. This enormous "living symphony" is precisely governed and coordinated, focused on the paramount task of maintaining the organism in its physically improbable living state. Governing and coordinating the reactions that enable the organism to stay alive is the function of the information that pervades the body. Information in this context is not a peripheral adjunct to biochemical processes but that which governs and coordinates those processes. The information that governs the organism is what differentiates one species from another, one individual in a species from others, and a healthy individual from a sick one. It also spells the difference between a normal cell and a cancerous cell, a healthy organ and a diseased organ.

It was believed that information in the body is limited to genetic information, and that genetic information is fixed for a lifetime. Current findings in biology and medicine indicate that this is not the case. The information that governs organic functions is more complex and comprehensive than the genetic code in the DNA, and it is not rigidly fixed but open to adaptation and modification. Even genetic information is modifiable. Although the sequence of genes in the DNA is fixed, the way that sequence affects the body is flexible: it is governed by the epigenetic system, and the epigenetic system is adaptive.

The way cells reproduce in the body is likewise modifiable: their operative program changes in interaction with the rest of the body. It now turns out that the information—the "programming"—of the cells can also be purposively modified. Recent discoveries in medicine (Biava 2009) demonstrate that stem cells in the body can be reprogrammed. If some of these cells are mutant—only replicating themselves—by reprogramming they can be reintegrated with the rest of the organism. When they are reprogrammed, cancer cells, for example, either die off (through apoptosis, programmed cell death) or become functional parts of the body. By replacing cancerous and degenerated cells with reprogrammed stem cells, many forms of cancer and neurodegenerative diseases can be eliminated. These previously fatal diseases become reversible maladies.

THE AKASHIC QI ATTRACTOR

According to the Akasha paradigm the information that coordinates the functions of a living organism is a specific pattern in the sea of A-dimensional information. This corpus of information governs action, interaction, and reaction throughout the manifest world. It also governs the functions of the living organism. It is a blueprint of normal organic functioning.

The blueprint of organic functioning for living species emerged in the course of the interaction between the M- and the A-dimensions. The A-dimension, as we said, in-forms systems in the M-dimension, and in-formed M-dimensional systems de-form the A-dimension. The information generated in this interaction is conserved in the A-dimension. The A-dimension is the memory of the M-dimension; it is the manifest world's "Akashic Record."

The sea of Akashic information includes the species-specific pattern that is the natural "attractor" of healthy functioning in the organism. This pattern results from the long-term interaction of a species with the A-dimension; it is the enduring memory of those interactions; and it codes the generic norms of viable species (see Sági 1998). For human beings it is the equivalent of the *Qi* (or *Chi* or *Ch'i*) of Chinese medicine, the *prana* of Hindu philosophy, and the *life energy* of the traditional Western healing arts. Without access to this Qi, prana, or life energy errors in cellular and organic interaction, reaction, and transcription would accumulate in the body and would lead to ever more serious and ultimately terminal malfunctions. This is inevitably the case—for biological organisms on this planet are inherently mortal—but the species-specific "Qi attractor" slows the degenerative processes and enables the organism to unfold the full potentials of its inherent vitality.

IMPLICATIONS FOR MEDICINE

Life can emerge and persist in the universe because living systems tune in to and resonate with information in the Akasha dimension. Malady and malfunction are errors in the way a living system receives and processes this information. In many cases these errors can be corrected. This has major and, in modern medicine, largely unexploited implications for the maintenance of health and the cure of disease.

In traditional societies people made more effective use of the Akashic Qi attractor in maintaining their health. Shamans, medicine men and women, and spiritual leaders were remarkably accomplished in safeguarding the physical condition of the people in their tribe, village, or community. Modern physicians, on the other hand, are more accomplished in curing diseases than safeguarding health. Their approach to disease relies on artificial means: the introduction of synthetic substances and surgical interventions. By these means modern medicine has prolonged life expectancy and produced treatments for numerous diseases. But synthetic substances and artificial interventions produce a plethora of unwanted and undesirable side effects. And they divert attention from the healing powers of natural substances and of closer harmony with natural rhythms and balances.

The methods of modern medicine have their place and utility, but they are not necessarily always the best way to maintain health and cure disease. Prior to the manifestation of disease there is a breakdown or blockage of information in the organism, and these conditions can be treated by reestablishing resonance with the organism's Akashic Qi attractor. Doing so is to treat the cause of the malfunction rather than its manifestation.

The Akasha paradigm suggests that the first task of the medical practitioner is to adapt the M-dimensional interactions of the organism for optimum conformance with A-dimensional information. This calls for attention to the relations of people to their social as well as natural surroundings. Stresses and strains in family and community impair the ability of the organism to cope with adverse conditions and toxic substances in its environment. They interfere with its conformance with Akashic information and thus diminish its vitality.

Patients can be helped to "tune in to" their natural environment. The organism is a psychosomatic system, in constant interaction with its surroundings. It is sensitive to information from the M-dimension as well as from the A-dimension. Both kinds of information are vital for health. In today's world there is an urgent need to regain contact with the A-dimension. The fuller our body's conformance with the species-specific Qi attractor, the more robust our health and the greater our capacity to resist toxic substances and negative influences.

There are practical methods for tuning our body for better conformance with the species-specific Qi attractor. Sophisticated devices measure the flow of energy and information in the organism. Some devices also correct the insufficient or blocked flows. Electronic systems such as the Nutri-Energetic Systems (NES) " human body field" scanners created by Peter Fraser and Henry Massey register energy flows throughout the body; Radionics, a system for analyzing body-field information (called "IDF," Intrinsic Data Field), attempts to reorganize the body's information flows.

The most widely used treatment employing information to rebalance flows in the body was pioneered over two hundred years ago by Samuel Hahnemann. In this method, called homeopathy, the remedial substance is highly diluted. In applications above the potency "D23" not a single molecule is likely to be present—yet in most cases the remedy proves effective.

Current developments based on Hahnemann's discovery include the "Psionic Medicine" practiced by the Laurence Society of Holistic Medicine in England (2000). The members of the society are reputed physicians who use a pendulum to reach a diagnosis and determine the remedy. This practice—not necessarily communicated to the patients— functions in the remote mode: only a "witness"—a sample of hair or a drop of blood—is needed to establish connection between healer and patient. Another information-based method is the "New Homeopathy" developed by the Viennese healer Erich Koerbler. Koerbler's method makes use of a specially designed dowsing rod to obtain information on the condition of the patient and to prescribe a remedy. The Hungarian healer Dr. Maria Sági has developed a sophisticated diagnostic and therapeutic system using the Koerbler rod. Her method works equally well in the proximity of the patient and at any distance. (See the exchange with Dr. Sági in chapter 13.)

Homeopathy, Psionic Medicine, and New Homeopathy are among the many currently developed—or just rediscovered—methods that identify flawed flows of information in the body and help rebalance them. Their curative power resides not in rebalancing one or another information flow, but in enabling people through the balanced flows to better tune into the human Qi-attractor.

The information that governs healthy organic functioning is available to all of us. When we access it, we can correct energy blockages, breakdowns, and malfunctions. The application of the Akasha paradigm to medicine recalls this traditionally well known but currently nearly forgotten method of organic self-healing.

10

Freedom

There is more to human freedom in the world than a science based on the old paradigm would have us believe. We are an organic part of a nonlocally interconnected universe, and we interact not only with its manifest dimension, but also with its Akasha dimension. This gives us a far greater degree of freedom than interaction with the manifest dimension alone.

There is no absolute freedom for any system in an interconnected world. Absolute freedom presupposes a total lack of ties to the rest of the world, and in this universe this is impossible. But absolute freedom is not only not possible, it is also not desirable. Freedom does not reside in being "free from" external influences, but in being "free to" act the way we wish to act in regard to them. In the latter sense we have a significant degree of freedom even in an interlinked and interacting world.

THE SCOPE OF HUMAN FREEDOM

Freedom in the world is neither nil nor full; it is a matter of degree. The scope of freedom is determined by external as well as internal factors. The external factors limit the scope of behavior. In regard to human beings they reduce the range of intended actions to the physically—and also psychologically and socially—feasible. The internal

61

factors are elements of freedom. They allow a living organism to select the way it acts from the range of possible ways. The relative weight of the external versus the internal factors differentiates between the freedom of an amoeba to move in relation to its food supply and the freedom of a human being to select the way he wants to live. For the amoeba the external factors are dominant, whereas for the human being the internal factors gain in importance. In complex biological systems the element of self-determination can be highly significant.

While in less evolved species information received from the external world is mainly in the form of an undifferentiated "feel" of the world, in the more evolved species the world is perceived through a rich flow of information that can be coupled with a wide range of responses. In a human being this flow is further differentiated as a series of articulated perceptions with conscious as well as subconscious, rational as well as emotive elements. This offers scope for a wide range of responses.

In the new paradigm we recognize information reaching us from the manifest as well as from the Akasha dimension. We select our response to the information that reaches us from both of these dimensions. We admit some of this information to our consciousness as bona fide perceptions of the world and exclude other information as irrelevant or illusory. In the modern world we exclude from consciousness most of the information that reaches us from the A-dimension. This constrains the scope of our response to the world around us; it limits the range of our freedom.

ENHANCING THE HUMAN POTENTIAL FOR FREEDOM

Like other living systems, we need to maintain ourselves in a dynamic state far from thermal and chemical equilibrium through the intake and processing of information, energy, and the quanta-based substances we regard as matter. This calls for constant high-level sensitivity to the vital flows of information, energy, and matter. Lest our free energy becomes depleted and our vitality impaired, we must select the right flows at the right time and couple them with the right responses.

The more complex the system, the more decisive is the selection of the information to which it responds, as well as the selection of its response to it. We achieve this "stimulus-response coupling" by processing the information we receive from the world. Our freedom is enhanced to the extent that this information is well processed: that the signals are properly selected, clearly differentiated, and accurately coupled with responses.

One aspect of our freedom is the purposive selection of the influences that act on us. Another aspect resides in the selection of our response. Whereas in comparatively simple organisms the responses to external stimuli are largely preprogrammed, in humans the response is conditioned by a series of "intervening variables." These are partially, but only partially, under our conscious control.

A vast array of sub- or nonconscious variables also determines our response to the information that reaches us. This array includes tacit preferences and unexamined values, cultural predispositions, and a range of acquired or inherited leanings, preconceptions, and prejudices. They shift the factors that determine our response to the world from the world to us. They highlight the crucial role of worldviews, values, and ethics, as elements of human self-determination and hence of freedom.

Consciousness can extend the range of our freedom. If we adopt consciously envisaged worldviews, and bring consciously envisaged goals and values to bear on our life, our freedom acquires an additional goal-oriented dimension. And if we allow not only the sensory information that connects us with the manifest world to penetrate to our consciousness but also the more subtle insights and intuitions that reach us from the A-dimension, we further extend the effective range of our freedom.

In addition to information that originates in the external world, we can respond also to information that we ourselves generate. As conscious beings capable of abstract thinking and imagination, we can envisage events, people, and conditions without actually experiencing them. We can respond to this self-generated information the same way we respond to information from the external world. We can recall the past and envisage the future. We are not limited to the here-and-now. Not only can we react, we can also proact.

This element of our freedom is vastly expanded by allowing the information that reaches us from the A-dimension to reach our consciousness. Akashic information is nonlocal information; it could have originated anywhere and at anytime and could concern anything in the world. Our contact with this information can take us beyond the here-and-now, to the everything-at-anytime.

11

The Good

We have the highest potential for freedom of any being on this planet. As conscious human beings we can be aware of this freedom and make purposive use of it. The question we address here concerns the humanly and morally best use of this freedom.

Morality enters this discourse because, if we can choose the way we act, we have the responsibility to choose it wisely. Evidently, we can act to maximize our own self-interest, and that is what most people do most of the time. But we can also act with a measure of altruism and public spirit. Acting in that way may not be contrary to our self-interest—at least not to our *enlightened* self-interest.

Self-interest makes us seek the satisfaction of our immediate desires and aspirations, but if our desires and aspirations are sound this is all right: then our desires and aspirations coincide. In a strongly interconnected and interacting world what is good for one is good also for the others. But what are the truly enlightened interests and aspirations?

THE GOOD

Philosophers have been debating what is truly good in the world for more than two thousand years. No definitive answer has emerged. In Western philosophy the view of the classical empiricists has prevailed:

judgments of good and bad are subjective; they cannot be decided unequivocally. At the most they can be related to what a given person, a given culture, or a given community holds to be good. But that, too, is subjective, even if it is subjective in relation to a group: then it is intersubjective.

In Akashic philosophy we can overcome this impasse: we can discover objective criteria for the good. These criteria do not carry the certainty of logic and mathematics, but they are more than subjective or intersubjective. They are as objective as any statement can be about the world. They refer to the conditions that ensure life and wellbeing in an interconnected and interacting universe. Enhancing these conditions is objectively good. These conditions can be briefly outlined.

As already noted, living organisms are complex systems in a state far from thermodynamic equilibrium. They need to meet stringent conditions for maintaining themselves in their physically improbable and inherently unstable condition. What is good for them is first of all to meet these conditions. Life is the highest value. But what does it take to ensure life for a complex organism on this planet? Describing all the things that this entails would fill volumes. But there are basic principles that apply to all living beings.

Every living system must ensure reliable access to the energy, matter, and information it needs to survive. This calls for fine-tuning all its parts to serve the common goal: to maintain the system as a living whole. The term *coherence* describes the basic feature of this requirement. A system consisting of finely tuned parts is a coherent system. Coherence means that every part in the system responds to every other part, compensating for deviations and reinforcing functional actions and relations. Seeking coherence for one's self is a truly sound aspiration; it is indubitably good for us.

But in an interconnected and interacting world the requirement for coherence does not stop at the individual. Living organisms need to be internally coherent, with regard to the fine-tuning of their parts, but they also need to be externally coherent, with well-tuned relations to other organisms. Hence viable organisms in the biosphere are both individually and collectively coherent. They are *supercoherent*. Supercoherence indicates the condition in which a system is coherent in itself and is coherently related to other systems.

The biosphere is a network of supercoherent systems. Any species, ecology, or individual that is not coherent in itself and is not coherently related to other species and ecologies is disadvantaged in its reproductive strategies. It becomes marginalized and ultimately dies out, eliminated by the merciless workings of natural selection.

The great exception to this rule is the human species. In the last few hundred years, and especially in the last decades, human societies have become progressively incoherent both with respect to each other and with their environment. They have become internally divisive and ecologically disruptive. Human societies could nevertheless maintain themselves and even increase their numbers because they compensate for their incoherence by artificial means: they make use of powerful technologies to balance the ills they have wrought. This, of course, has its limits. Whereas in the past these limits appeared mainly on the local level, today they surface also on the global scale. Species are dying out, diversity in the planet's ecosystems is diminishing, the climate is changing, and the conditions for healthy living are reduced. The system of humanity on the planet is nearing the outer bounds of sustainability.

We can now say what is truly good in this crucial epoch. It is to regain our internal and external coherence: our supercoherence. This is not a utopian aspiration, it can be achieved. But it calls for major changes in the way we think and act.

AWAKENING TO "THE GOOD"

Striving effectively to supercoherence requires more than finding technological solutions to patch up the problems created by our incoherence. It requires reconnecting with a mind-set that traditional cultures possessed but modern societies have lost. This is a mind-set based on a deep sense of oneness with each other and with nature.

In today's world many people feel separate from each other and from the world. Young people call it *dualism*. The prevalence of dualism has grave consequences. People who feel separate tend to be self-centered and egoistic; they do not feel connected with others and do not feel responsibility for them. Behavior inspired by this sense of duality creates tooth-and-claw competition, eruptions of mindless violence and anger, and the irresponsible degradation of the living environment. This mind-set has dominated the modern world, but there are signs that it is losing its grip on individuals and societies.

Ever more people, especially young people, are rediscovering their oneness with each other and with the world. They are rediscovering the power of love—rediscovering that love is more than the desire for sexual union, that it is a profound sense of belonging to each other and to the cosmos. This rediscovery is timely, and it is not mere fantasy: it has its roots in our holographically whole, nonlocally interconnected universe.

Love is the way to supercoherence. Achieving it is health enhancing and socially and ecologically sound. It gives rise to behaviors and aspirations that are good for us, good for others, and good for the world. Supercoherence is objectively good. It is the highest value philosophers called "The Good."

PART FOUR

Questions, Answers, and Reflections

A new paradigm in science has implications well beyond the bounds of science; they encompass all aspects of human life and aspiration. We have explored some of these aspects and implications in part three. Now in part four we embark on a still more wide-ranging exploration. We enter in dialogue with and listen to the comments of creative thought leaders in philosophy, science, the media, the healing arts, and practical activism.

12

On the Meaning of the New Paradigm

From an In-Depth Dialogue with
David William Gibbons

Historian, Writer, and Cofounder of
Universal One Broadcasting

David William Gibbons (D.W.G.): I would like to talk today about our times and the transition that we collectively experience. Perhaps, indeed, even the decade that has led to our times today. You have written many books on this subject and have many projects in hand. What is your very basic definition of this transition, and how we can fully witness it?

Ervin Laszlo (E.L.): I think we are at the end of an epoch, an epoch based on a mistaken consciousness. This epoch—which began several hundred years ago—is an aberration in the history of humankind, in which we have tried to use our emerging powers, physical powers, to manipulate the world to be our own and for our own immediate interests. Thereby we have gradually and increasingly subverted its balances, its own direction. Human manipulation of

70

the world has become really powerful, with the combined power of technology and the power of enterprise, both of which have even taken over from local and national politics. It is now the global enterprises that in a very real sense rule the world, served by a consumer mentality. This is therefore an aberration in the consciousness that guides humanity through history.

This epoch has come to a head in the last few decades, particularly in the last decade and certainly since the end of 2012. We are now witnessing a leap, a transition, a transformation toward a consciousness, an epoch, a culture, and a civilization where people are more in harmony with one another and with our planetary environment.

D.W.G.: You have called it *The Dawn of the Akashic Age* in your book of that title (Laszlo and Dennis 2013). There you talk about the Neolithic Era, some ten thousand years ago, as the basis of a human civilization, which first developed a type of hubris perhaps. We could call this an ancient form of consumerism. Throughout the series of deep dialogues I have been conducting, I have been attributing the emergence of this hubris, and the beginning of modern consumerism, particularly in the United States of America, to the postwar years. But you say that in actual fact this consumerism, this need for power through financial wealth, has been with us for ten thousand years. What are your thoughts on this?

E.L.: I wouldn't necessarily say that the new form of consumerism is the factor that triggers all of our problems. What emerged ten thousand years ago in the Neolithic period, in the Levant, in what used to be called the Fertile Crescent, was a belief. Actually, it was a mistaken belief, the belief that humanity is above and beyond nature. We can domesticate animals as we can domesticate plants, but we cannot domesticate nature. However, on the basis of this mistaken belief, our ancestors began to fit their environing nature to their needs. Later Francis Bacon articulated this "hubris" when he said that our task is to extract the secrets of nature from her bosom, to use for our own benefit. Yet rather than fitting nature to our needs, we need to fit ourselves to nature—and by this I mean the whole web of life on the planet.

D.W.G.: In the work I am undertaking I find myself talking increasingly to members of generations that are set in old ways. Certainly I consider myself in some respects to be placed broadly speaking within that group, those who remember well how entering into the seventies we had the Sinclair calculator, and the story of course unfolds from there. We were hardwired in a way and brought into being through a highly materialistic world. When I talk with youngsters (graduates from Berkeley, London, and other universities), I find that they are experiencing a similar but opposing effect, taken from different perspectives and in different times. It's almost as if they're creating a balance between two generations. How can that work? How can we shape that condition, so that both generations may actively work together in finding that balance?

E.L.: Balance will emerge if we allow the processes of change to unfold. Change does not come from the center. It does not come from the dominant layers or from the established generation. Change comes from the outside, from the periphery. We know that this happens in nature when an ecosystem with multiple populations becomes unbalanced. Darwin thought that then the dominant species mutates to adapt to the new circumstances. Since the 1980s we have known that dominant species do not change. No species changes by itself. Certainly the random mutations that Darwin thought were the engine of evolution cannot be the explanation, because there are too many possibilities for change. The search space of the genetic endowment of a species is far too large for random change to have a reasonable chance of coming up with a viable species. What happens is that the dominant species, the mainstream, dies out because it is no longer capable of maintaining itself and the system that it has dominated. Then space is opened for change to come from the outside, from the periphery. Change is occurring there all the time.

Even in society change is coming from the periphery, from alternative cultures and "mutant" individuals. You and I are change agents; everybody who thinks like you and I are what biologists call a *hopeful monster*. These are people who have mutated before the time has come, before the environment has changed in such a way that they could take over as the dominant species. Some of these hopeful monsters will take over, and some may not. For the latter, they will remain just hopeful. In any event and in the current environment they are monsters. But they may not be that when the world is changing around them. They may then become the standard, the norm.

So I don't see the need, or even the real possibility to intervene or to mediate between the old and the new, between the established culture/generation and the emergent. We must allow the emergent to emerge. When it does it is going to be a revolution. It will be a bloodless revolution because ultimately and basically it will be a conceptual revolution—a worldview revolution, a values revolution. A revolution of what we do because of what we realize we should be doing. There are alternative ways of thinking and acting available, but not to everybody. The alternatives are not there for the one and a half billion people who live below the poverty line according to World Bank estimates. They must just survive somehow. But anybody above the poverty line has a choice in regard to consumer behavior, political behavior, social and cultural behavior.

The biggest factor lies with those who are financially comfortable enough to have such choices. If they include active opinion makers then they can create ideas, ideals, and values that can spread. They can influence others by how they themselves behave, as consumers, as political activists, and as citizens, through what they support and aspire to. What counts is that they are open and creative. But those who rule the world today do not belong to this category.

If they don't change, they are condemned to extinction. They will either take the system down with them or quit the scene in time to give space to the young in spirit, the creative, the imaginative, allowing them to move to center stage and create a world in which all can live.

D.W.G.: One could look back at the rise and fall of the Roman Empire or of the British Empire to find examples of radical change. Indeed we could look at other examples of collapse. In any event it may not always be apparent at the time. It tends to be insidious, I believe, particularly if you're living through it. You don't necessarily see or are conscious of the changes because you're in the midst of them. But there are different levels and different parts of this journey to travel through and I suspect it will be over the next ten or twenty years that much will manifest and become clear.

In many ways it would be good to accomplish this without a complete systems collapse. The question is whether that is entirely possible, as systems are so reliant upon existing consumerism trends and financial systems. We need to be proactive in understanding what to expect and how to deal with these factors. But could it be that over the next five, ten, or fifteen years that there will be a collapse. Is it a collapse that we need to be preparing for?

E.L.: A collapse will be coming almost certainly within the next five, ten, or fifteen years. But whether it will be a global collapse, whether it will be an irreversible collapse, that remains to be seen. If we can have a perceived crisis rather than a lived crisis, then we could begin to change without having to undergo a collapse. In this way, when we start feeling that the roof is falling in, we don't actually have to suffer through it. There is a spontaneous gut reaction warning us that this will be happening. I don't believe that we have to predict the future. Certain trends, when they enter into a critical stage, create a spontaneous reaction within us. We begin to feel an approaching crisis. People begin to sense that something dramatic is about to happen. This is what we call euphemistically a collective instinct. An individual as well as a collective survival instinct is something real, which has to do with what the world is like, with the interconnected nature of the world.

What is now happening in the world is "feel-able" and it is being felt. It is not by chance that there is widespread social and political unrest, local ecological breakdowns, and even technological catastrophes. There is a growing perception of the need for change and of the real possibility of change, except for the diehards who just want to preserve the old system.

There are still some who think you cannot change the world or human nature. But except for these people, there is a deep-seated movement of change, bringing a growing willingness to change. There is a movement to reconnect us with a larger reality—with the global reality of the web of life on this planet. The web of life is a whole system that is subtly but effectively acting on and influencing its members. This idea, of a whole that acts on its parts, is known in science as downward causation. Biologists discovered it decades ago. Not only do the parts of a system influence the whole, which is upward causation, the whole also influences all its parts.

For example, our brain has consciousness as a whole and this consciousness influences the way its neurons work. Something like this is occurring now on the planetary level. The system of humankind is responding to the crisis that it is about to experience. This is propelling humanity into a new epoch, a different age. The current age is on its last legs. It doesn't work anymore, and it can't get us anywhere any longer. This is felt by sensitive people, and especially by young people, which makes them hopeful: monsters today, but thought leaders tomorrow.

D.W.G.: Yes, they are the hopeful monsters. I refer to this development as "riding the dragon." I suppose we all travel through different octaves and different experiences, whether part of an older or younger generation. Devoid of ego, is there not really a time when you have to speak your truth and be on the forefront in driving this transition through? That may be vague but one has to be assertive at times, even while being and acknowledging that you're part of the whole and inseparable. Your thoughts on this?

E.L.: If you have a message that can encapsulate the gist of the problem, then you are morally obliged to speak up, to broadcast the message, especially if you feel that the way people receive the message is still largely subliminal and unconscious. Given that people feel the coming transformation in the world but cannot articulate it, the best way to broadcast your message is to live it: to "be the change" as Gandhi said. Then the message will spread through osmosis and by empathy. If people sense that you have changed, they will begin to change in a similar way themselves.

There are different degrees of efficacy and obligation to broadcast the message, but broadcasting it without changing oneself is meaningless. Only when you yourself have changed can you motivate others to change. The root factor is the evolution of consciousness. You don't "make" other people evolve their consciousness. You don't "teach" the new consciousness. People discover it on their own, with the help of the change in your own consciousness.

By an evolved consciousness I mean a new mentality, a new set of values, a recognition of our ties to each other. Helping people evolve this new mentality is the task of the genuine teacher, a real Guru, or spiritual Master. It is a very different task from that of a dictator or even the director of a company, even though enlightened business people recognize that people have to evolve their own consciousness and not just obey instructions in evolving it.

A sustainable and humane world can only be a democracy, but the problem with democracy is that the people must rule, and they must have the wisdom to rule. They must see the situation without bias by politics and self-interest.

Here a metaphor helps: the spaceship. We are the crew of a natural spaceship in orbit around the Sun. The way we now run

Supply +
Demand

this spaceship is unsustainable: we are using up the power in its batteries. I mean the deposit of the Earth's fossil fuels. We are also using up the available material resources, the mineral and biological resources. At the same time we are accumulating waste and junk on this spaceship. If we continue doing this eventually we will suffocate and will not have enough resources to live. It's very important to have this kind of metaphor in mind, because it offers us a true picture of the situation as it exists on a planetary scale.

We live on a natural spaceship. With solar and solar-based energy we have nearly infinite energy resources. But we are making use of only a small fraction of them. We're still not realizing that using enduringly available flow energies and recycling material resources is an absolute precondition of our collective survival. We need to acknowledge the need to become part of the world. We have opted out of the world, mistakenly thinking that we are above and beyond it. Now we either get back into it or pay the consequences—with our collective existence.

D.W.G.: What is the principal motive, our principal intention, when we're connecting with other people in order to find this collective existence and to insure that we travel through this transition with some form of progress? Because surely it does begin and end with that soul and heart connection? It has to start there beyond anything else.

E.L.: We are all connected, intrinsically and permanently connected. That is the new paradigm, the Akasha paradigm, emerging at the cutting edge of the sciences. We only disregard this new insight at our own risk. If we can open up our mind and our heart to our oneness in the world we will come up with the solution. The precondition for this is to allow the wisdom that's in us to become operative. This wisdom has guided people through the ages. It has been expressed in symbolic form as prophetic insight, which was then often solidified as written dogma. Thus religious doctrines became sectarian and divisive, fragmentary rather than universal. Yet the basis of all the wisdom in the world's cultural traditions is our connection and oneness. How do we act on the basis of this wisdom?

The only way we can do this is by acting together at a deep level. By sensing our oneness, by cooperating, by becoming coherent. We are no longer coherent either with each other or with the world around us. Traditional societies possessed coherence: they were wholes, even if they fought with each other. Sometimes they were violent. But they sustained themselves for thousands and thousands of years, and this was because they had a basic coherence. This coherence has been broken apart in our materialist, fragmented modern world. We each just want to do our own thing. The world out there is a jungle. We are responsible only for ourselves; all others are strangers, sometimes allies but more often enemies. They certainly are not us—we and they are two things. This is duality, the opposite of oneness. We need to find our way back to what we really are, an intrinsic part of the whole, which is humanity as a whole, which is the web of life as a whole. We need to become coherent again with ourselves, and with others around us. If we do that, if we move in that direction, we are a positive agent of change in this crisis-bound world.

D.W.G.: Would cohesiveness share parallels with collaborativeness? I mentioned Walter Russell, the rhythmic balance, and Sophia—the feminine that found a hubris in herself, in the aftermath of which the masculine dominated for hundreds or thousands of years to our present day. I do think that this is an important point as we travel toward the end of the journey, particularly for those who are trying to figure out the relationship between the male and the female. What is that balance? What is that nurturing? What does it take to bring back the feminine not only in the feminine, but also in the masculine, to realize our potential for finding unity?

E.L.: What does it take? I'd say, to reconnect, to realize that we are one, that there is a larger system of which we are a part. We are part of a series of larger wholes, wholes within wholes. What it takes is to recover the intuitive feeling that we are a part of it, that we are connected. So I could say, coming right down to it: it takes love, the deep, embracing feeling of love. Love is the recognition that the other is not other. The other is also me and I'm the other. The world is not beyond or outside of me; it's inside me just the same way as I'm inside the world. There are no absolute boundaries between me and what I see as the world. There are only different gradients of intensity in our relationships.

I may be more closely related to my sons and partners in life than to someone I have never met on the other side of the world. But I'm related to all of them; it's a difference only in intensity. Ultimately I am connected with everyone, just as I'm connected to the person closest to me. If I love the person closest to me then I also love all other people because we are all part of the same whole—we are part of each other.

This is the key to what we should look for when we are asking which way we must go. I don't see even the remotest possibility of creating a sustainable and flourishing world on this planet unless we embrace this embracing love. Is this idealistic; is it utopian? Normally, the answer would be yes. But it is not utopian in a period of critical instability, of impending crises. In this period many things are possible, except maintaining the status quo and going back to the past. Embracing all-embracing love is neither the one nor the other. We could see it perhaps as going back to some local groups of people who had developed coherence based on their love for one another. But this has never happened for humankind as a whole. Yet now it must happen, because we have become a planetary species. We must extend the embracing love that members of families felt for each other to all people on the planet. Doing this is not an option but a necessity. I believe it is feasible. We can connect with each other now in so many ways; we can be aware of our shared history and know that we have a shared future. We can discover that we are all one family. Utopia becomes a possibility at this critical juncture of our history.

D.W.G.: The secrets of the Divine Feminine, the power of the feminine and return to balance—what would you say is important about acknowledging and understanding this concept?

E.L.: Cohesion, cooperation, empathy, caring are all feminine values. We need them in the world and should see to it that these values have an active role in shaping the world. This means that those who hold these values must have more of a say in this world. The way this world is structured now it is based on masculine values. It is power oriented, short term, self-centered, oriented to accumulating wealth as an expression of power, to making others do what we want. These are typically masculine values. They come from the past of human communities when the males went out hunting and the females stayed home to take care of the hearth, of the family. Today there is a basic need to care for the family, for the community. Yet the world we have created is based on the hunter mentality, on violence mentality, on power mentality. We must inject into this world more caring values, more typically feminine values. I don't mean the values of women—I mean values that typically women hold more than men. Women can hold masculine values too. In fact most of the women who are successful in today's world in terms of money and power act on the basis of masculine values. They're successful at business and in politics for that very reason. But typically women hold values that are more feminine than these. We need more typically feminine values in the world. They must balance the predominance of the masculine values.

D.W.G.: We're at the end of our journey for today. Why do you do what you do?

E.L.: It's not a rational motivation. You see that I live out in the country, in pleasant surroundings. I could just stay at home and enjoy life. But I don't do that—I go out and look for meaning in what I am doing. If I spend a day without having either a positive thought or some exchange with others that I consider positive, I think of it as a wasted day. There are so many things I must still do.

D.W.G.: So I should not feel concerned about the fact that creating and crafting these depth dialogues is what I was meant to do and that this is what I love to do.

E.L.: You are doing something that moves the world in the right direction. This is an evolving world, a world that is inherently coherence-seeking, a world that seeks higher and higher levels of unity and oneness. Either we are a part of this movement in the world or we opt out of it. We can even go against it. That's the freedom of a human being. But if we combine our freedom with a sense of responsibility, for ourselves, for others, and for nature, then we go with the evolution of the world. That is the embracing, loving way to go, the satisfactory way—for me at least.

D.W.G.: Ervin Laszlo, thank you so very much.

13

Healing through the A-Dimension

An Exchange with Dr. Maria Sági

Psychologist, Healer, and Science Director
of the Club of Budapest

Ervin Laszlo (E.L.): Healing with information rather than with bio-chemical substances and invasive methods is the "soft" approach to medicine—not an alternative to mainstream medicine, but an evermore evident basic complement to it. I became convinced of the possibility of such healing as a logical inference from the tenets of the Akasha paradigm, and this conviction was greatly reinforced by the experience I had with the healing method you practice. Even before I came across the possibility of healing with information, whether proximally or at a distance, you had discovered that such healing works. We came from different starting points and arrived at the same destination. Your approach is important, because it provides a practical, experiential confirmation of the soundness of the principles that constitute the Akasha paradigm.

Healing, I have noted in chapter 9 of this book, should be possible through the Akashic or A-dimension. You are practicing a form of healing that, it seems to me, uses precisely this kind of information. I would like to explore with you how this works. Allow me

82

to ask you some questions. The first thing I would like to know is how you discovered that there is another way of healing than the standard method of Western medicine.

Dr. Maria Sági (M.S.): I had a life- and mind-transforming experience at the time I was a young research associate working in the field of the psychology of music and art, and it was this experience that made me enter on the path to healing. Out of friendship and intellectual curiosity I accompanied a friend of mine, the wife of a long-time colleague, who had some health complaints that resisted standard medical advice. She went to see an old priest in the country who had a reputation for healing in ways that went beyond the usual practice of modern medicine. The priest, Pater Louis, worked with a pendulum. When he finished checking my friend, he spoke to me, "Dear child, you are perfectly healthy, only garlic is poison for you." At that moment Pater Louis was standing behind me with his pendulum in his hand. I had not asked him to examine me, yet what he found spontaneously was entirely correct. How did he know that I could not stand garlic, and all foods that had garlic in them? I had this intolerance for as long as I can remember. I asked Pater Louis what kind of diet he would suggest for me. He told me to avoid meat, milk, bread, and sugar. This, I found later, is precisely the macrobiotic diet. I adopted it, and my life changed dramatically. I had more energy, more vitality, more robust health. I became a devoted disciple of this wonderful old priest.

I began to read up on macrobiotic and related Eastern healing literature. I had a good basis for healing, since I had four years of university study in medicine. I began to study various types of phytotherapy and medical dowsing. Then I went to Amsterdam to the Kushi Institute to receive training as a teacher and healer in macrobiotics and continued my study of the works of Rudolf Steiner. I set out on a path that was to take me to a twofold career: social science, and alternative healing.

E.L.: This life- and mind-transforming experience started you out on your second career, but how did you continue? How did you acquire the knowledge and the skills you now possess?

M.S.: Further surprising things happened to me. Four years after meeting Pater Louis, my father died. At that time he was living in Austria and I did not have everyday contact with him. Yet after he died I had remarkable experiences, very depressing and worrying. Many things became clear to me about my relation with my father. This confirmed for me that beyond the world we experience there is another, deeper dimension. Talking with my friends and doing research on my own, I came to understand that in this dimension everything is recorded and everything can still be experienced. This helped me to understand not only the experiences I had following the death of my father, but also how Pater Louis could treat the people who consulted him whether they were next to him or far away. Because he did treat people from a distance: if someone could not come to see him in person, he or she would send a photo and Pater Louis would use his pendulum on the photo just as he would have on the person in front of him. It worked just as well.

The next milestone on my way to becoming a healer came when I had the good fortune to meet the Austrian scientist and healer Erich Koerbler. He developed a healing method called New Homeopathy. Koerbler diagnosed the condition of his patients according to the principles of Chinese medicine, using a specially designed dowsing rod that oscillates and indicates the condition of the patient. The eight different movements of the rod enabled Koerbler to obtain a precise and detailed picture of the energy state of his patient. He discovered that his method works through the electromagnetic (EM) field. With the help of his one-arm dowsing rod he measured the electromagnetic radiation emanating from the body of his patients. He experimented constantly and discovered that certain geometrical forms function as "antennas" in the patient's EM field. These forms affect the body and can correct flawed information. They can produce a healthier condition in the body. Koerbler's "vector system" situates the movements of the dowsing rod within a system of coordinates. Observing the movements of the rod provides indications of the compatibility or noncompatibility of a given substance or other input with the patient's organism. Substances and inputs that are compatible with the healthy functioning of the organism are indicated by one type of movement, various degrees of less than beneficial to seriously harmful inputs are indicated by a different set of precisely defined movements. The movement of the rod indicates also the cause of a harmful input, and how serious it is.

I worked with Koerbler for three years, and after his unexpected death I continued to teach his method in Hungary, as well as in Germany, Switzerland, Austria, and Japan.

E.L.: Erich Koerbler's method is remarkable and, as you found, remarkably effective. Yet it is a local method; it requires that the patient be next to the healer. But if I am not mistaken, this method, like the pendulum-based method of Pater Louis, can also be used at a distance. Can you tell me if that is true, and if it is, how it works?

M.S.: Let me give you some background on how I came to practice the Koerbler method both proximally and at a distance.

News of the effectiveness of Koerbler's method continued to spread, and after Koerbler's death more and more people turned to me for help. Many of them lived in foreign countries and could not come to see me in person. I discovered that I could treat them also remotely, using their photograph.

Then I became acquainted with the healing method practiced by the physicians who are members of the Psionic Medical Society in England. They use a pendulum to diagnose their clients and homeopathic remedies to effect a cure. A special chart against which they note the movement of the pendulum gives them the key to the diagnosis as well as to the cure. This works similarly to the vector system of Erich Koerbler, but these doctors treat their patients only at a distance. They obtain a protein sample from their patients—a drop of blood, a few strands of hair, for example—and with that sample they get the same results as they would by examining their clients in person. I worked with the Psionic Medical Society for nine years, and during this time I acquired a thorough knowledge of their method.

In the years that followed I elaborated the technique of remote healing on my own, using Koerbler's geometrical forms combined with the remote healing method of the Psionic Medical Society. I applied this combined method when I was face to face with my patients, as well as when I only had their photograph in front of me. Once I examined my patients in person I could also dispense with their photo: I could focus on them entirely in my mind.

E.L.: You say that when you contact your patients you receive information on their bodily state, and that this information comes to you through the Akasha dimension. Could it not come from the patient himself or herself directly?

M.S.: If it were information coming directly from the body of the patient it would be information on his or her current condition. But I can receive information on the condition of the patient from anytime in his or her lifetime—even from just after birth, and sometimes from before that, from the period of gestation in the womb. I can concentrate on any period I like of my patient's life and observe the movement of the Koerbler dowsing rod. In this way I can locate the period that is immediately pertinent to the health problem of my patient—because most health issues have roots in something that has happened to us in our life. Often I can verify the occurrence of an event that created the health problem through independent sources—for example, by the mother of the patient, or another person who had witnessed the given happening. Then I try to correct the negative effect of that information by applying the healing forms discovered by Koerbler and developed by myself. If I am successful the symptoms of the problem disappear quite shortly (for example, neurodermitis, chronic coughing, and the common cold symptoms), or this may take a few weeks.

Proceeding in this way I treat the cause of the problem and not its consequences. Treating the consequences—which is what standard medical practice does—calls for local information, information on the current state of the patient. But treating the cause requires nonlocal information, and that can only come from the A-dimension.

E.L.: What is your understanding of what is happening when you treat your patients at a distance, which, as we now see, can be distance in space and also in time? How can the information you send actually work on the patient's body?

M.S.: As you have written in your books, the energy field of the individual is in constant interaction with the Akasha dimension. The organism, like all quanta and multiquanta systems, is embedded in that plenum similarly to vessels in the sea. There is constant communication between the receptive networks of the brain and the wavefields of the plenum.

The communication of generations and generations of people with the A-dimension creates the species-specific pattern. Within that generic pattern of humankind there is also the specific pattern of an individual. This is the individual's "morphodynamic pattern" (Sági 1998). In my view every individual has his or her own morphodynamic pattern. This pattern is inclusive: it encodes all the events that affect the individual, including the behavior of the neural nets that underlie his or her consciousness. It encodes the characteristics of the physical body on the one hand and the characteristics of mind and consciousness on the other. The morphodynamic patterns of the individual are the Akashic Qi attractor for that individual.

The individual is healthy and resilient as long as his or her bodily state conforms to the norms of the Qi attractor. Every deviation of the organic state from the norms of the Qi attractor means a weakening of life energy. It is a prelude to illness. If it is not corrected it is followed by the onset of disease.

E.L.: How does the species-specific morphodynamic pattern—the Akashic Qi attractor—produce health and healing for the patient?

M.S.: When the healer sends healing information to the patient, he or she reinforces the match between the patient's body and the Qi attractor. Thanks to this reinforced match the patient's immune system becomes better able to maintain bodily functions within the limits of normalcy—in a condition of vitality and health.

E.L.: Can this nonlocal way of diagnosing a person's condition and healing his or her problems be learned? Can it be practiced by anybody, or does it call for being a qualified physician or healer?

M.S.: As you know, accessing another person's body and mind through the Akasha dimension is possible for everyone. But to heal through this contact calls for sound knowledge on the part of the physician or healer. I can access the energy and information field of my patients by focusing on them, but I cannot heal them with any assurance unless I am thoroughly acquainted with the nature of their health problem. Only then can I suggest the rebalancing required to produce healing. I need to have a sound method at my disposal for making a qualified diagnosis. And that calls for mastering the use of a healing system, such as, for example, the vector system developed by Erich Koerbler.

E.L.: Are you personally, emotionally involved with the people you heal? Is that a necessary condition of healing in this way?

M.S.: My mind must be clear and I must have an honest wish to heal. But I must not be unduly involved with those I treat: I must distance myself in order to receive unbiased information. I need to be open in regard to the information I receive about the nature of the problem, the nature of the remedy, and, in the case of homeopathy, also about the posology and potency of the required remedy. Only if I am open to receiving unbiased Akashic information, and also have the knowledge to apply it properly, can I heal reliably and effectively.

E.L.: What does your healing experience tell you about the connection that exists between you and your patients—and about the world in which such connections can come about?

M.S. Practicing this method is a great satisfaction for me. Day after day and year after year I am amazed at the results and grateful for being able to achieve them. This for me is clear proof that we are part of a larger whole; that there is a subtle connection between all of us—a connection of which we can make active use. It is evidence that not only do we receive information from the world, but that we also shape the world, even with our mind and consciousness. And that we can shape it responsibly, maintaining our own health, the health of others, and helping to heal the discord and disconnection that besets this world.

14

What Is the Akasha?

A Commonsense Q & A with Györgyi Szabo

Akasha Paradigm Researcher and Former Program Director
of the Ervin Laszlo Center for Advanced Study

Györgyi Szabo (G.S.): When all is said and done, just what is the Akasha? And why is it important that I should know it?

Ervin Laszlo (E.L.): The Akasha is a dimension in the universe that subtends all the things that exist in it. It not only subtends all things: it generates and interconnects all things, and it conserves the information they have generated. It is the womb of the world, the network of the world, and the memory of the world.

This new discovery—more exactly, new rediscovery of an ancient insight—is important for science and it is important for you. It is important for science because it enables the integration of the various theories of the natural disciplines into what Einstein called the simplest possible and yet comprehensive scheme—a scheme that conveys a meaningful picture of the world based on the accessible evidence we have of the world. And it is important for you because a recognition of the Akasha dimension and its relation to the world we observe and live in can orient your thinking and guide your steps as you head toward the great transformation that will change not only your picture of the world, but also your world.

G.S.: How does the Akashic worldview differ from our commonsense picture of the world?

E.L.: The new paradigm gives us a very different concept from that which most people hold in the modern world. Understanding the world through the lens of this paradigm calls for a veritable "gestalt switch." We normally think of the things we experience as real, and the space that embeds them as empty and passive, a mere abstraction. We need to turn this around. It is the space that embeds things that is real, and the things that disport themselves in space that are secondary.

This concept emerges from the findings of cutting-edge physics. Space, quantum physicists realize, is not empty and passive; it is a filled and active plenum, even though physicists still refer to it as the "vacuum." In the emerging view space is the ground, and the things we know as real things in the world are the figures on the ground. They are not just figures on a ground; they are figures *of* the ground. The things we consider real are manifestations of space—more exactly, manifestations of the Akasha, the cosmic matrix that fills space.

There is a good metaphor for this concept of the world. Think of waves traveling over the surface of the sea. When you look at the surface you see waves moving toward the shore, waves spreading out behind ships, waves interfering with waves. The waves move from one point on the sea toward another, yet there is nothing in the sea that moves that way: the molecules of water on the surface do not move from one place to another, they just move up and down. The motion of the waves is an illusion—an illusion not in the sense that there is nothing that corresponds to it, but in that it is not what it appears to be. The waves travel across the surface of the sea, but the water of the sea does not travel. The same applies to the motion of things in space. Things do not move across or over space, they move in, more exactly within, space. They are conveyed by space.

The view this gives us is very different from the commonsense view. The real world is not an arena of separate things moving across intervening space; it is a manifestation of a cosmic matrix. All things are part of that matrix and are conveyed in and by the matrix. Illusion is not the bare existence of things, but their separateness. All things are in and of the matrix and in the final count are one with the matrix.

G.S.: Can we be certain that this is the correct view of the world?

E.L.: This is an important question, and I am glad that you asked it. This question is usually asked by skeptics: people who want to disprove a statement or theory. But it also should be asked by those who are ready to believe in it. The plain fact is that we cannot be *absolutely* certain that the new paradigm gives us the correct view of the world, but we can be *reasonably* certain that it does. There is no absolute certainty in science beyond the formulas of logic and mathematics. It is only there that we can have proof of the truth of our conclusions, because the proof, the same as our whole chain of reasoning, is "axiomatic": it is defined in its own terms without reference to anything else.

Einstein pointed out that as far as the proofs of mathematics do not refer to reality they are certain; and as far as they refer to reality they are not certain. Abstract schemes can be certain, but they become uncertain when they are applied to the real world. Already two-and-a-half thousand years ago Plato warned us that our ideas about the world are at best a likely story. At this point in the evolution of our insight into the nature of things, the Akasha paradigm is, I believe, the likeliest story.

G.S.: Then this is a spiritual view—or is it a scientific view?

E.L.: I am not surprised you frame this question in this way. In today's world science and spirituality are distinct, even opposing, positions. If you are spiritual you cannot be scientific; and if you are scientific, you are very likely not spiritual. But the new paradigm overcomes this specious divide. You can be both. This in itself is not new. Genuine spirituality has always been based on the recognition of a deeper intelligence at work in the cosmos. The prophets and teachers of the world's religions gave their own interpretation of this intelligence, identifying it in ways that accorded with the concepts and language of their time. But a literal interpretation of their scriptures would be a mistake, for it would suggest that only their particular interpretation is true and valid.

This would be similar to the dogmatic claim in mainstream science that only the data conveyed by the senses gives true information on the world, while anything beyond that is idle speculation: "metaphysics." A mature science recognizes that the world is far greater and deeper than our sensory experience of it, just as a mature religion recognizes that the higher or deeper intelligence its doctrines suggest is the real core of the cosmos. A mature science is spiritual, and a mature religion is scientific. They are built on the same experience, and they reach basically the same conclusion.

G.S.: Is the Akasha the intelligence of the world?

E.L.: The Akasha is indeed a kind of intelligence. In the spiritual context we could call it the consciousness or intelligence of the world, and in the scientific context we can best view it as the logic or "program" of the world. It is what makes the world intelligible, makes the stars and planets and the atoms and organisms behave in a way we can comprehend. The Akasha paradigm conveys insights that were traditionally in the domain of spirituality and religion.

We can envisage the A-dimension as a kind of divine intelligence. It is immanent in the world, an intrinsic part of the cosmos. But in our immediate experience as human beings that intelligence is transcendent: it is beyond "our" world.

G.S.: Which of these views is the correct one?

E.L.: Both the "immanentist" and the "transcendentalist" views of a divine intelligence are correct. Their validity depends on how we approach this intelligence. "Sub specie eternitatis," viewed from a vantage point from outside our world, we envisage the cosmos in its totality. In that perspective the A-dimension is equivalent to the divine consciousness that permeates the world. But in a perspective from within the world, more exactly within our human experience of the world, the A-dimension is not an immanent spirit suffusing the cosmos but an infinitely available space- and time-transcending field of information.

G.S.: Does this mean that all people can experience the divine intelligence in the Akasha?

E.L.: The intelligence encoded in the Akasha informs all things in the manifest world. It in-forms our brain and body, and it in-forms our mind and consciousness.

Akashic information in-forms our brain and mind, but not necessarily our conscious mind. It is often, and in the modern world usually, repressed from waking consciousness. If we are not to repress it or dismiss it as fantasy and imagination, we must experience it consciously. And for this we need to enter an altered state of consciousness.

This was not always necessary. Shamans and medicine men and women, prophets and spiritual teachers have been able to access this cosmic intelligence as part of their everyday experience. They were frequently in altered states of consciousness; they cultivated them consciously, by rhythmic dancing, chanting, drumming, psychedelic herbs, rites and rituals, and disciplined meditation among other things. But in the modern world we have been so thoroughly convinced that everything we experience must come through our senses that we do not seek experiences that do not involve sensory apprehensions. And if such experiences occur spontaneously, we dismiss them as illusion or fantasy. The exceptions to this rule are artists, spiritual people, and creative individuals in all walks of life, even in science. Often their key insights come in altered states, in prayer and in meditation, in esthetic experience, or in dreams, daydreams, hypnagogic states between sleep and wakefulness, or in deep introspection.

When we enter an altered state of consciousness images, ideas, and intuitions flow into our consciousness that transcend the range of our sensory perceptions. These elements are part of the totality of the information in the cosmic matrix: the Akasha. This information is in a distributed form, as in a hologram. That means that all elements of the information are present in every part of it. In delving deeply into our consciousness, we access a holographic fragment of this information—of the information that "in-forms" the universe. In a modest but real sense, we "read the Akashic Record"— the record of all the things that are, and have ever been, in the whole world.

G.S.: This is a mind-boggling prospect. But let us go to the bottom line: What do we know now that we did not know before?

E.L.: This is, indeed, the bottom line. In the final count, science is about understanding: knowing the world and all things in the world. A science based on a more adequate paradigm should convey more understanding, deeper knowledge, than a science based on the old.

How does the Akasha paradigm do in this regard? What knowledge does it give us that the previous paradigm did not and could not give? I will try to give a clear answer to this fundamental question.

In his 2009 book devoted to querying the "true nature of the universe," biologist Robert Lanza, of Wake Forest University, writing with Bob Berman created a kind of report card he called "Classic Science's Answers to Basic Questions" (Lanza with Berman 2009). According to Lanza's report card, science is failing: it gives no knowledge at all in regard to eleven of thirteen basic questions, and it conveys a tentative answer to the twelfth and a negative answer to the thirteenth.

Would a science based on the new paradigm do better? Here we contrast Lanza's questions and the answers he gives for "classic science" with the answers we can now offer on the basis of the new paradigm.

Q. How did the Big Bang happen?

A. Classic science: Unknown. **Akasha paradigm science:** Through the action of the repulsive forces that arose when the collapsing prior universe reached quantum dimensions.

Q. What was the Big Bang?

A. Classic science: Unknown. **Akasha paradigm science:** A phase transition between universes, or phases of the multiverse; not a Big Bang, but a Big "Bounce."

Q. What, if anything, existed before the Big Bang?

A. Classic science: Unknown. **Akasha paradigm science:** A preceding universe, or phase of the multiverse, with physical properties similar to that of our own.

Q. What is the nature of dark energy, the dominant entity of the cosmos?

A. Classic science: Unknown. **Akasha paradigm science:** Not yet known, but offering orientation for the search, as the answer is likely to be in the constitution of the hidden dimension, the physically real basis of spacetime and background of matter and energy in the universe.

Q. What is the nature of dark matter, the second most prevalent entity?

A. **Classic science:** Unknown. **Akasha paradigm science:** The same as above—not yet known, but offering orientation for the search, as the answer is likely to be in the constitution of the Akasha, the physically real basis of spacetime and background of matter and energy in the universe.

Q. How did life arise?

A. **Classic science:** Unknown. **Akasha paradigm science:** The processes we see as basic to life originated as coherent relations emerged in time in the rich welter of organic molecules on the watery surface of some satellites in orbit around active stars.

Q. How did consciousness arise?

A. **Classic science:** Unknown. **Akasha paradigm science:** Consciousness did not "arise"—it has always been present as an aspect of the intrinsically psychophysical universe.

Q. What is the nature of consciousness?

A. **Classic science:** Unknown. **Akasha paradigm science:** It is the mental, more exactly "mindlike," aspect of the matterlike systems in the manifest dimension of the universe; a display or reflection of the information contained in the universe's A-dimension.

Q. What is the fate of the universe; for example, will it keep expanding?

A. **Classic science:** Seemingly yes. **Akasha paradigm science:** With a high probability it will reach a balance between the initial force of expansion and the gravitational force of contraction, after which it will recontract to quantum dimension—to reemerge as the next universe (or next phase of the multiverse).

Q. Why are the [physical] constants the way they are?

A. **Classic science:** Unknown. **Akasha paradigm science:** The values of the constants have been progressively evolved in the preceding cycles of the multiverse and were passed on to our universe in the phase transition we recognize as a Big Bounce.

Q. Why are there exactly four forces?

A. Classic science: Unknown. **Akasha paradigm science:** The universal forces are not limited to four: the force (or field) responsible for nonlocal interaction is as basic and universal as the four classical fields, and there are various quantum fields and forces as well.

Q. Is life further experienced after one's body dies?

A. Classic science: Unknown. **Akasha paradigm science:** After the body dies life appears to persist as a form of consciousness in the Akasha, and can be experienced and communicated with, as shown in near-death experiences, after-death communications, and medium-transmitted contact (Laszlo and Peake 2014).

Q. Which book provides the best answers?

A. Classic science: There is no single book. **Akasha paradigm science:** The classic science answer is correct—there is no single book, nor is there ever likely to be one, although some books offer better answers than others today, and very likely also in the future.

15

Comments on the Akasha Paradigm by Leading Scientists and Thinkers

EDGAR MITCHELL

Astronaut

In the very ancient past the natural human desire to understand our world, its content, and the interactions in nature was stimulated by exploring beyond the local environment, discovering new flora and fauna and perhaps another tribe of humans living in a locale different from one's own. New words, new thoughts, and new relationships presented themselves to mind and needed description and discussion. Experiencing nature and new beings expanded our understanding and allowed the creation of beliefs about how all life fits together in the larger scheme of things.

Looking back from the modern present we can chronicle the arising of many languages and many cultural beliefs in and about our world. The ancient mystics and wise men in every culture were the leaders in prescribing the nature of reality as they saw it, and the rules and procedures that should be followed in interacting with each other and with nature. Local cultural beliefs turned into local religions that claimed to speak "the Truth" about nature and the cosmos, including stories of the origins of all things and rules of behavior humans needed to follow for successful lives and a successful social order.

As long as we remained local, such tribal procedures, beliefs, and rules sufficed to regulate and inform our societies. The beginning of distant travel and interaction with remote peoples more often brought conflict and strife than peace and harmony in the short run. In the Middle Ages kingships, empires, and travels to different continents became common in the Western world. The Christian religion dominated Europe and set the standard for thinking and social interaction in the emerging nations. Disagreeing with the Church could be labeled as heresy and bring on death by burning at the stake. Then along came the philosopher, mathematician, and thinker René Descartes. He wrote that body and mind, the physical and the spiritual, belonged to different realms of reality that did not naturally interact. The Church authorities accepted this idea, which allowed Europe's intellectuals free thinking as long as they avoided the subject of mind and consciousness, the province of the theologians.

Shortly thereafter, Sir Isaac Newton published his laws of motion. Modern science and the classical laws of physics were born. Newton's theory was based on the conviction that experiment and testing were necessary to prove its application to reality. Interactions in nature had to be measured and proven with mathematical precision. This classical period in Newtonian physics lasted for four hundred years, until the turn of the twentieth century, when Albert Einstein changed our understanding of the nature of space and time, and Max Planck, Erwin Schrödinger, and Paul Dirac brought in quantum physics as a necessary element for our understanding of interactions at the subatomic level.

The particular rules and the basic beliefs people used to understand the nature of reality constitute what philosopher Thomas Kuhn called a *paradigm*. Descartes, followed by Newton and his laws of physics, initiated a four-hundred-plus-year reign in which classical physics was the

paradigm for science. The discovery of the quantum world and the codi-
fication of its interactions in the 1920s ushered in changes that launched
a paradigm shift that took up most of the twentieth century. However,
the two revolutionary theories, general relativity and nonlocal quantum
mechanics, have not been successfully unified, and the post-Newtonian
paradigm lacked integral consistency. It has become clear, on the other
hand, that the Cartesian separation of matter and mind is a faulty concept: both matter and mind are basic elements of reality. Research on consciousness surfaced by the end of the twentieth century as an important
topic in science after four centuries of neglect. Today, the understanding
of reality wrought by the rise of quantum theory and the accumulation
of evidence regarding the active role of consciousness in the world cata-
lyze a basic paradigm shift. This is what Laszlo's new book is all about.
What it accomplishes merits our serious and urgent attention.

Bringing to the surface science's dominant paradigm and examining its power—or its failure—to provide an integral and realistic understanding of the world is one of the basic accomplishments of Laszlo's new
book. The second is to outline the new paradigm that could overcome
the shortcomings of the old and endow the theories of the contemporary sciences with integral meaning and realism. Science has outgrown
its dominant paradigm. We need a new paradigm to understand the
world that emerges at the cutting edge of the sciences. Laszlo's Akasha
paradigm fulfills this requirement. This is a revolutionary accomplish-
ment of enormous scope and indisputable relevance. We need to take
account of it, discuss it, and follow it up with sustained research.

DAVID LOYE
Author, Founder of the Darwin Project

Behind Ervin Laszlo's brilliant development of the new paradigm lies
the long history of efforts to understand who we are, where we've been,
and what happens next for us in evolution.

Among the ancient Hindus and Mayans, and over and over again in history, what emerges is the picture of the clash of the vast, over-riding mind-sets we call paradigms. For a long time one paradigm will rigidly hold most of us captive, then out of a time of cataclysmic trouble and confusion—such as we're experiencing now—a new paradigm will emerge.

Where Laszlo's Akasha paradigm fits into this picture is with the great shift in our time toward a bold new perspective that embraces, then weds and transcends both the old paradigms of religion and the paradigms of mainstream modern science.

With the Akasha paradigm we are entering what might be called an Age of Reconciliation. Rooted in a vast understanding of the heady dynamics of advanced physics, the Akasha paradigm weds the best of religion and the best of science into a powerful expression of a new partnership between progressive science and progressive religion to beat back the forces that are trying to drive us backward in evolution. Laszlo's grounding for the kind of mind-set we need to gain a better future has been examined and blessed by more leading scientists than any other I know of.

KINGSLEY DENNIS

Sociologist, Writer, and Cofounder of WorldShift International

At each stage of human evolution we are confronted by phenomena that call for us to investigate our paradigms of knowledge and understanding. At each stage we are asked to rise to the challenge to conceptualize, visualize, and vocalize these insights that are necessary to compel us forward along the unfolding of social evolution. How we navigate our present and potential future(s) is fundamental to how we survive as a species codwelling on a vibrant and life-sustaining planet. It is clear that our incumbent frames of scientific knowledge are in need of timely readjustment. Ervin Laszlo's presentation of the new paradigm does exactly that.

The Akasha paradigm returns our way of thinking to an integral consciousness, a nonlinear mode of understanding that prompts us to accept the reality of nonlocal interactions. This view shifts us away from the dominant mental-rational worldview toward a perspective that fosters an ecological-reciprocal relationship as we exist within an inclusive whole. This calls for nothing less than a mutation in human consciousness.

By adopting the Akasha paradigm we are bringing greater meaning into our life. This paradigm does not destroy, or collapse, our current or older models; rather, it updates our previous stages of knowledge into a more inclusive model that better serves to explain how the manifest world exists—and can exist—within a "hidden dimension" that underlies the structure of a more complete, inclusive energetic reality. The Akasha paradigm allows for the existence of materiality while simultaneously embracing the unification of all known phenomena—it is a unity paradigm at heart, and as such it is life sustaining in its implications for humankind.

The Akasha paradigm as set forth by Ervin Laszlo tells us that life exists within a pattern that is coherent beyond our wildest dreams. Living systems seek harmony throughout their evolutionary journey. This insight uplifts our spirit and imbues our life with new meaning.

DAVID LORIMER

Author, Lecturer, and Educator

In his recent book, *A New Science of the Paranormal,* Lawrence LeShan (2009) provides a trenchant analysis of the reaction of scientific materialism to paranormal phenomena. He points out that impossible events do not happen, so, by definition, if an event has happened it is possible. Events require explanations in terms of theories. So if the theory is incapable of explaining the fact, then so much the worse for the theory, rather than fact. It is the theory that needs revising, so that the fact is reframed.

The history of psychical research and parapsychology has provided many examples of scientists trying to explain away facts that they regard as a priori impossible or what C. D. Broad called "antecedently improbable." This may be a sophisticated turn of phrase, but it disguises a presupposition or prejudice. The bedrock of a worldview is a series of presuppositions or assumptions about the nature of reality. This is the topic of Rupert Sheldrake's brilliant book *The Science Delusion* (2012). In this book he discusses a number of scientific dogmas and turns them into questions. These include the propositions that nature is essentially mechanical, that matter is unconscious, that brains produce consciousness, that memories are stored as material traces in the brain, that minds are confined to the head, and that unexplained phenomena like telepathy are illusory.

None of these propositions is true, as Sheldrake convincingly demonstrates, yet they are dogmatically adhered to despite more than one hundred years of evidence to the contrary. This shackles the spirit of inquiry and stifles the real progress that might be made if mainstream scientists had more courage in questioning these dogmas, defying the peer pressure of their colleagues. As Nikola Tesla, another neglected genius, put it: "The day science begins to study nonphysical phenomena, it will make more progress in one decade than in all the previous centuries of its existence." Of course, a great deal of research has been carried out, but virtually none of it is acknowledged or incorporated into the mainstream.

The reason for this is the stranglehold of the ideology of scientific materialism that assumes that consciousness is a by-product of physical processes in the brain and cannot act nonlocally. This also means that telepathy and out-of-body experiences are impossible in principle, and that death spells the extinction of consciousness and personality. The standard approach of scientific materialism to parapsychology is to question the integrity of the experimenters or the rigor of the experiments. Dean Radin and others have convincingly demonstrated that these phenomena do occur and therefore require an explanation. Many people are resistant to paranormal evidence because they lack a coherent theory.

This is where Ervin Laszlo's Akasha paradigm comes in and provides a new framework for understanding consciousness and its operation. New theories should adequately explain the data that they address. Rather than continuing to defend scientific materialism against all comers, scientists should free up that spirit of open enquiry that is the true nature of the scientific process. As I pointed out at the beginning, if a theory is incapable of explaining well-attested evidence, then that theory needs expanding or replacing.

The time has come for a true paradigm shift in our understanding of reality to take account of coherence and nonlocality at all levels. It is not sufficient to account only for local rather than nonlocal interactions. As Laszlo observes, nonlocality is in fact a basic feature of the universe. He postulates Akasha as the hidden dimension of the universe, one that in-forms everything in the manifest physical world, but which is also an element of consciousness as experienced from the inside. Our embeddedness in nonlocal coherence gives us access to information beyond the reach of the physical senses. Thus paranormal phenomena are not unexpected but rather to be anticipated in this extended framework.

Among the forerunners of these ideas were William James and Henri Bergson with their concepts of radical empiricism and creative evolution and their understanding that the brain tends to filter out the nonphysical, especially among those with the left hemisphere dominance. They would have no difficulty, along with Alfred North Whitehead, in recognizing that we are an intrinsic element in "a locally and nonlocally interconnected and interacting universe." This dynamic rather than deterministic order also leaves room for self-determination, which Laszlo regards as the essence of freedom.

This is a refreshing prospect consistent with emerging findings in physics, biology, psychology, and parapsychology. We should cast off the shackles of scientific materialism while remaining rigorous in our approach to explaining the full range of human experience—not only normal and abnormal, but also exceptional capacities. What we currently call the paranormal or supernormal should in fact be regarded as normal within this wider Akashic framework.

STANLEY KRIPPNER
Anthropologist and Humanistic Psychologist

The standard cosmology of the twentieth century told a story starting with the Big Bang, a single event that created the cosmos. Eons later, gigantic swirls developed that became galaxies. Eventually, at least one solar system emerged that contained a planet capable of developing and sustaining life.

This story supplanted earlier stories that attributed creation to deities (or a single deity), along with ingenious mechanisms such as epicycles that had been invented to preserve the notion that Earth was the center of its solar system. When these stories unraveled, it was not an easy matter to tell another one. One early storyteller paid with his life while another was put under permanent house arrest. The twentieth-century storytellers look back at the earlier stories as superstitious tales, taking pride in the ascent of science as a guide to constructing creation tales based on reasoned logic, empirical observation, and carefully constructed experiments. However, it was these very tools that revealed that the twentieth-century story had its own epicycles, mechanisms that did not hold up once quantum mechanics entered the picture.

In this brilliant book, Ervin Laszlo tells a story for the twenty-first century, a narrative inspired by a much earlier story from ancient India, the Akashic Records, perhaps the first "theory of everything" that highlighted nonlocality, a term absent from the twentieth-century standard cosmology but paramount a century later. When physicists observed that separated subatomic particles at the microlevel continued to interact at impossible distances, this phenomenon was dismissed as an interesting but unimportant quirk. In any event, it could not be repeated at the macrolevel.

However, the data began to accumulate from physics, chemistry, biology, and the "disreputable" field of parapsychology that nonlocal events spanned the gap between micro and macro. Time, space, and the ephemeral construct called "consciousness" engaged in a dance that quantum physicist David Bohm called the "holomovement." Indeed, Bohm was one of the first to provide a narrative that was dismissed at the time but that probably insured his place in the history of the philosophy of science.

Drawing upon Bohm, Schrödinger, Einstein, and many others, Laszlo's story tells us that the Akashic world is a locally as well as a nonlocally interconnected and interacting world. The Akashic world includes a dimension in the universe that subtends all the things that exist in it. It not only subtends all things, it generates and interconnects all things.

Some people of a particular doctrinaire bias will read this story and will assume that this dimension supports such fairy tales as Creation Science or Intelligent Design. Far from it. The Akashic cosmos is a self-organizing whole that creates itself. It is reminiscent of Charles Darwin's closing statement in *The Origin of Species*. Darwin ended his story by commenting, "It is interesting to contemplate a tangled bank, clothed with many plants of many kinds, with birds singing on the bushes, with various insects flitting about, and with worms, crawling through the damp earth, and to reflect that these elaborately constructed forms, so different from each other, and dependent upon each other in so complex a manner, have all been produced by laws acting around us."

This is as close as Darwin could come, given his era and his knowledge, to describing nonlocality. But he was on target when he exclaimed, "There is grandeur in this view of life." And in his remarkable update of Bruno, Copernicus, Galileo, Darwin, Einstein, and others, Ervin Laszlo, heeding Plato's warning that even the best-designed account of the nature of reality is but a likely story, has now good reason to tell us that the Akasha paradigm is the likeliest story we can tell today.

DEEPAK CHOPRA

Physician and International Thought Leader

The Akasha is not the electromagnetic field or a physical field but literally a realm of transcendental consciousness, the consciousness from which the whole universe arises and into which the universe again subsides.

Understanding Laszlo's work on the Akasha paradigm gives insight into platonic values like truth, goodness, beauty, harmony, and ambition. These give rise to spontaneous and not imposed morality, coming from an experience of our higher self. They give scientific insight into the most fundamental nature of the universe. At the fundamental level the universe is more than space, time, and energy, spin and charge, and all the things physicists talk about. The Akasha is the fundamental building block of an evolving, maturing mind. Understanding it opens the way to a higher morality and a higher consciousness.

KEN WILBER

Philosopher, Writer, and Founder of the Integral Institute

Ervin Laszlo's recent book is highly welcomed for many reasons. First, his basic proposition—that there is a transuniversal holofield operating in a largely unmanifest dimension that gives unity to all things manifest, and is the actual ground from which the entire manifest realm emerges and to which it returns, and is ultimately responsible for the coherence of the universe itself—is well known to the mystical traditions the world over. (In Integral Theory, it is the causal dimension, consisting of what the Lankavatara Sutra calls *vasanas,* or a storehouse memory of every thing and event that has ever occurred, and which in turn gives rise to subtle phenomena, which give rise to gross or physical phenomena, in a whole series of "stepped-down" versions of their causal origin.)

But what makes Laszlo's version of this causal field—which he appropriately connects with the ancient doctrine of the Akashic record—is his extensive discussion of this field in light of recent "hardcore" developments in the conventional sciences. You might say he gives a very thorough third-person account of a spiritual reality that can also be experienced, as he acknowledges, in first-person terms in meditation, and in second-person terms in I-Thou intuitions and grace. Laszlo lists over a dozen inadequacies in present-day sciences that absolutely scream out for a new paradigm, a paradigm that is holistic, connective, holographic, supraluminal, nonlocal, and transuniversal (surviving multiverse to multiverse events). This is no New Age loopy speculation, but thoughtful, careful, fully outlined forays into the most respected fields of science available, at their very leading edges.

Laszlo is that incredibly rare individual, somebody who is literally as at home in scientific as in spiritual domains. As he indicates, the more mature science becomes, the more spiritual it becomes; and the more mature spirituality becomes, the more scientific it becomes. The centuries-old battle between the two is simply, absolutely, radically outdated. The same unifying principles necessary to account for a soul's union with God are necessary to account for the coherence and synchronization of the human body, or M-theory, or cosmology and biology, in themselves. Laszlo knows this, and this is his great genius: that every scientific hypothesis he advances has been checked against a spiritual background—and vice versa. This is exactly the type of thinking that is necessary to usher us into a truly Integral Age, and Laszlo is one of its great pioneers. This book is highly recommended for professional and layperson alike, and it is even a good place to start as one picks up Laszlo's other, equally significant books.

The Akasha Paradigm in Science

The two appendices to this work provide documentation and support for the Akasha paradigm in science.

Appendix I reviews the observations and experiments that provide the evidence that unseat the previous and still dominant paradigm and open the door to exploring the nonlocal-interconnection-based Akasha paradigm.*

Appendix II puts forward two mathematically elaborated hypotheses that show that phenomena in space and time can be accounted for in reference to a dimension in the universe that is beyond space and time. They contribute to laying the foundations for the Akasha paradigm in the technical domain of the new physics.

---

*This material represents a summation and updating of topics covered in comprehensive detail in my previous Inner Traditions books: *Science and the Akashic Field* (2004, 2007), *Science and the Reenchantment of the Cosmos* (2006), *Quantum Shift in the Global Brain* (2008), and *The Akashic Experience* (2009).

Nonlocality and Interconnection

A Review of the Evidence

Let us recall Einstein's definition of science: it is the human endeavor that seeks the simplest consistent scheme of thought that can tie together the observed facts. The Akasha paradigm enables scientists to tie together more of the observed facts than the previous paradigm. It encompasses facts that were anomalous for the latter, facts that testify to nonlocal connection throughout the range of scale and complexity in nature.

In the three sections of this appendix we review the facts that serve as evidence—even if necessarily just partial and preliminary evidence—for the new paradigm. The fields where these observations come to light are wide and encompassing. They include (i) the world of the quantum, (ii) the living world, and (iii) the universe in its largest dimensions.

NONLOCALITY AND INTERCONNECTION IN THE QUANTUM WORLD

Quanta, the smallest known entities of the physical world, do not behave like ordinary objects. Until an instrument or an act of observation registers them, they have neither a unique location nor a unique state. And they are nonlocally interconnected throughout space and time.

The Weird World of the Quantum

The quantum state is defined by the wavefunction that encodes the superposition of all the potential states a given quantum can occupy. The superposed state of the quantum is the pristine state, in the absence of all interaction. The duration of the pristine state may vary. It can be just the millisecond that it takes a pion to decay into two photons, or it can be ten thousand years in the decay of a uranium atom. Whatever its length, it is the state of superposition known as one tick of a quantum clock, or q-tick. According to the Copenhagen interpretation of quantum theory, reality (or in any case space and time) does not exist during a q-tick, only at the end of it, when the wavefunction has collapsed and the quantum has transited from the superposed indeterminate to the classical determinate state.

It is not clear what brings about the collapse of the wavefunction. Eugene Wigner speculated that it is due to the act of observation: the consciousness of the observer interacts with the particle. Yet it has turned out that the instrument through which the observation is made can also impart the crucial impetus: the wavefunction collapses whether or not an observer is present.

Another aspect of the "weirdness" of the quantum world is becoming dispelled: the curious limitation stated in Heisenberg's uncertainty principle. Heisenberg's celebrated principle tells us that all the properties of the quantum state cannot be measured at the same time: when one property is measured, a related property becomes unmeasurable—it becomes blurred, and its value may go to infinity. This, however, has now been shown to not necessarily be the case. Experiments begun in 2011 at Canada's National Research Council, and reported by physicists at the University of Rochester and the University of Ottawa in early 2013, have shown that it is possible to measure some key related variables of the quantum state (the "conjugate" variables) at the same time. Their new apparatus measures the quantum state responsible for one of the variables in such a weak way that the measurement does not significantly disturb the quantum state. In that case the second of the conjugate variables can likewise be measured, since the information received from the aspect of the quantum state corresponding to that variable remains valid.*

Whereas some of the weirdness imputed to the quantum world in the twentieth century resists clarification, the entanglement of phenomena at the quantum level becomes more and more evident. The quantum world is entangled in space and, as we shall see, it has now been shown to be entangled in time as well. The world at the quantum level is entirely nonlocal.

The Famous Nonlocality Experiments

Nonlocality in the world of the quantum surfaced in a series of experiments, each more anomalous for the prequantum classical paradigm than the other.

The series of experiments started with the explorations of Thomas Young in the early nineteenth century. Young made coherent light pass through an intervening surface with two slits. He placed a screen behind this surface to receive the light that penetrated the slits and found a wave-interference pattern on the screen. This seemed to suggest that photons, as waves, had passed through both slits. But how could this happen when the light source is so weak that only one photon is emitted at a time? A single packet of light energy should behave as a corpuscular entity; it should be able to pass through only one of the slits. Yet even in this case an interference pattern builds up on the screen. Can even a single photon behave as a wave?

John Wheeler carried out an entire series of more precise and sophisticated experiments in the 1970s and '80s. In Wheeler's experiments photons are emitted one at a time and are made to travel from

---

*The June 25, 2013, issue of *Nature* reported that Paul Busch of the University of York came up with experimental proof of some aspects of the Heisenberg uncertainty principle. But he recognized at the same time that there are other theories and experiments, including those of Masanao Ozawa, now at Nagoya University in Japan, that indicate that the principle can be violated.

the emitting gun to a detector. The latter clicks when a photon strikes it. A half-silvered mirror is inserted along the photon's path; this splits the beam, giving rise to the probability that one in every two photons passes through the mirror and one in every two is deflected by it. To confirm this probability, photon counters that click when hit by a photon are placed both behind this mirror and at right angles to it. The expectation is that on the average one in two photons will travel by one route and the other by the second route. This is confirmed by the results: the two counters register a roughly equal number of clicks—and hence of photons. When a second mirror is inserted in the path of the photons that were undeflected by the first, one would still expect to hear an equal number of clicks at the two counters: the individually emitted photons would merely have exchanged destinations. But this expectation is not borne out by the experiment. Only one of the two counters clicks, never the other. All the photons arrive at one and the same destination.

It appears that the singly emitted and therefore supposedly corpuscular photons interfere with one another as waves. Above one of the mirrors their interference is destructive—the phase difference between the photons is one hundred eighty degrees—so that the photon waves cancel each other. Below the other mirror the interference is constructive: the wave phase of the photons is the same, and as a consequence they reinforce one another.

Photons that interfere with each other when emitted moments apart in the laboratory also interfere with each other when emitted in nature at considerable intervals of time. The "cosmological" version of Wheeler's experiment bears witness to this. In this experiment the photons are emitted not by an artificial light source, but by a distant star. In one experiment the photons of the light beam emitted by the double quasar known as 0957 + 516A,B were tested. This distant quasi-stellar object is believed to be one star rather than two, the double image being due to the deflection of its light by an intervening galaxy situated about one-fourth of the distance from Earth. (The presence of a galaxy, the same as of every mass, is believed to curve the spacetime matrix, and so curve the path of the light beams that propagate through it.) The deflection due to this "gravitational lens" action is large enough to bring together two light rays emitted billions of years ago. Because of the additional distance traveled by the ray deflected by the intervening galaxy, it has been on the way fifty thousand years longer than the ray that traveled the direct route. Even though they originated billions of years ago and arrive with an interval of fifty thousand years, the two light rays interfere with each other just as if the photons that constitute them had been emitted seconds apart in the laboratory. It appears that whether light particles are emitted at intervals of a few seconds in the laboratory, or at intervals of thousands of years in the universe, those that originated from the same source create wave-interference patterns with each other.

The interference of photons and other quanta is extremely fragile: any coupling with another system destroys it. More recent experiments came up with a still more astonishing finding. It turned out that when any part of the experimental apparatus is coupled with the source of the photons, the fringes that indicate interference among them vanish. The photons behave as classical particles.

Further experiments have been devised to determine through which of the slits a given photon is passing. Here a so-called "which-path detector" is coupled to the emitting source. The result is that as soon as the apparatus is in place, the interference fringes weaken and ultimately vanish. The process can be calibrated: the higher the power of the which-path detector, the more of the fringes disappear (Wheeler 1984).

An even more surprising factor appears as well. In some experiments the interference fringes disappear as soon as the detector apparatus is readied—and even when the apparatus is not turned on. In Leonard Mandel's optical-interference experiments two beams of laser light are generated and allowed to interfere. When a detector is present that enables the path of the light to be determined, the fringes disappear. But they disappear regardless of whether or not the determination is actually carried out. It appears that the very possibility of "which-path detection" collapses the photons' wavefunction: it destroys their superposed state (Mandel 1991). This finding was confirmed in experiments carried out at the University of Konstanz by Dürr and collaborators in 1998 (Dürr, Nonn, and Rempe 1998). In these experiments the puzzling interference fringes were produced by the diffraction of a beam of cold atoms by standing waves of light. When there is no attempt to detect which path the atoms are taking, the interferometer displays fringes of high contrast. However, when information is encoded within the atoms as to the path they take, the fringes vanish. Yet the instrument itself cannot be the cause of the wave function's collapse—it does not deliver a sufficient "momentum kick" since the back action path of the detector is four orders of magnitude smaller than the separation of the interference fringes. In any case, for the interference pattern to disappear, the labeling of the paths does not need to be actually carried out: it is enough that the atoms are labeled so that the path they take can be determined. It appears that the measuring apparatus is "entangled" with the object that is measured.

Entanglement occurs not only across space; it appears that it also occurs across time. Observational evidence for this often-voiced but hitherto speculative hypothesis came about in the spring of 2013, in experiments performed at the Racah Institute of Physics at the Hebrew University of Jerusalem. Physicists Megidish, Halevy, Sachem, Dvir, Dovrat, and Eisenberg coded a photon in a specific quantum state, then destroyed that photon; as a result, as far as may be ascertained, a photon in that particular quantum state no longer existed. Then they coded another photon for that particular quantum state. They found that the state of the second particle was instantly entangled with the state of the first even though the latter no longer existed.

This demonstrates a surprising fact: entanglement can obtain between particles that have never existed at the same time. How can this be? The physicists speculated that the state of the first photon must have been stored somehow in spacetime (Megidish, Halevy, Sachem, Dvir, Dovrat, and Eisenberg 2013). The phenomenon of spatial entanglement suggests the presence of a spacetime medium that conveys the state of quanta over finite distances. The phenomenon of temporal entanglement reinforces this hypothesis. There is now experimentally produced observational evidence that spacetime is an instantaneously interconnecting medium with memory.

The EPR Experiment

The historically first demonstration that quanta that at one time existed in the same quantum state remain entangled over finite distances came about in response to a thought experiment put forward by Albert Einstein with Boris Podolski and Nathan Rosen in 1935. The Einstein-Podolski-Rosen or EPR experiment was said to be a "thought experiment" because at the time it could not be physically tested.

Einstein and colleagues suggested that we take two particles in a so-called singlet state, where their spins cancel out each other and yield a total spin of zero. We split the particles and project the two halves a finite distance. Then we should be able to measure one spin state on one of the halves, and another spin state on the other. If so, we would know both states at the same time. Einstein believed that this would show that the limitation specified by Heisenberg's uncertainty principle can be overcome.

The Heisenberg principle tells us that the more accurately we can specify one of the parameters in the quantum state of a particle, such as momentum or spin, the less accurately we can specify its other parameters—for example, its location in space. When we fully define one parameter, the others become entirely blurred. (As already noted, according to the indeterminacy principle it is not possible to measure all parameters of a particle's quantum state at the same time.)

Einstein believed that this interdiction is not intrinsic to nature; it is due to our systems of observation and measurement. The EPR thought experiment was suggested to show that this is indeed the case.

When experimental apparatus sophisticated enough to test the experiment came online, it turned out that the Heisenberg principle holds up. Indeed, it holds up under conditions that Einstein did not envisage; namely, over any finite distance.

Scores of experiments have now been carried out over ever-greater distances. They testify that there is an instant, intrinsic interconnection between the particles. Separation in space and time does not divide particles that originated in the same quantum state, regardless of how long ago they shared that state and how far they have been projected from each other since then. It is not even necessary that the particles should ever have coexisted at the same time. The fact that a particle is in a quantum state shared by another particle, whether past or present, appears sufficient to create instant interconnection between them. It is in this sense that we can say the quantum world is intrinsically and wholly nonlocal.

NONLOCALITY AND INTERCONNECTION IN THE LIVING WORLD

Nonlocal Connections in the Organism

More than half a century ago Erwin Schrödinger, convinced of the nonlocality of the quantum state, suggested that nonlocality need not be limited to the quantum world. The kind of order manifested in living systems, he said, is not order of a mechanical kind: it is dynamic order. Dynamic order is not an order based on chance encounters among mechanically related parts, and it cannot arise by random collisions among individual molecules. It is an order based on ultrarapid systemwide "nonlocal" connection among all elements of the system, even among those that are not contiguous with each other.

Schrödinger's insight is corroborated by the finding that tissue in living systems constitutes so-called Bose-Einstein condensates. (These condensates, first postulated by Satyendra Bose and Albert Einstein in 1924, constitute a dilute gas of bosons cooled to temperatures close to absolute zero. Under these conditions a large fraction of the bosons occupy the lowest quantum state where quantum effects become apparent.) It was only seven decades later, in 1995, that experiments could demonstrate the presence of the condensates in the organism. The citation for awarding the 2001 Nobel Prize in physics to Eric A. Cornell, Wolfgang Ketterle, and Carl E. Wieman stated "for the achievement of Bose-Einstein condensation in dilute gases of alkali atoms, and for early fundamental studies of the properties of the condensates." Cornell, Ketterle, and Wieman have shown that supercooled aggregates of matter—in their experiments rubidium or sodium atoms were used— behave as nonlocal waves. They penetrate throughout the condensate and form interference patterns. Information within Bose-Einstein condensates is transferred instantly, and this produces the kind of coherence previously associated only with lasers and quantum systems.

Instant nonlocal connection among all parts of the organism is a precondition of maintaining the organism in the living state. Such connection suggests that in some respects the living organism is a macroscopic quantum system. In quantum systems molecular assemblies, whether neighboring or distant, resonate in phase: the same wavefunction applies to them. Attractive or repulsive forces are generated depending on the phase relations among the wavefunctions, and faster and slower reactions take place in regions where the wavefunctions coincide. Through such processes long-range correlations come about that are nonlinear, quasi-instant, heterogeneous, and multidimensional. While molecular reactions at different points carry out the individual functions, the coordination of the functions occurs by means of nonlocality: through the quasi-instant multidimensional transfer of information among the assemblies.

Nonlocal Connection among Parts of the Organism: Pathbreaking Observations

In the 1960s lie-detector expert Cleve Backster placed the electrodes of his lie detector (a polygraph) on the leaves of a plant in his office. To his surprise, he found that the instrument registered reactions by the plant that correlated with his own experiences. For example, the polygraph revealed marked deviations in the electrical resistance of the plant at the very instant when Backster nearly had an accident on the street below. The correlation was maintained even when the leaf was detached from the plant and trimmed to electrode size, or shredded and redistributed between the surfaces of the electrodes.

These findings have been replicated in subsequent experiments. Ben Bending of the University of California at Los Angeles reported that his results "support the claim that plants have some sort of perceptual faculties that allow them to sense human emotion" (Bending 2012). Although the signals manifest electrically, the connection may not be electromagnetic, given that Faraday cages and lead shielding fail to block the signals. According to Bending this suggests the possibility of a nonlocal connection between plants and their caretakers.

Evidence for nonlocal connection among different parts of one and the same organism was discovered when Backster (1968) took oral leukocytes (white cells) from the mouth of test subjects and moved them to locations that ranged from five yards to over eight miles from the subject. He monitored the cells' net electrical potentials and fed the signals to a chart drive unit that provided a continuous recording of the changes. He then stimulated his subjects with visual images designed to evoke an emotional response and observed the variations in the cells' electrical potentials. It turned out that the changes in the potentials were correlated in time, amplitude, and duration with the emotional responses of the subjects.

In one experiment a young man was handed an issue of *Playboy* magazine. When he came to the centerfold photo, which was a nude picture of actress Bo Derek, his EEG manifested an emotional response that lasted during the whole time he viewed the picture. Changes in the electrical potentials of his distant cells mirrored the changes in his emotional state. When he closed the magazine the values of his responses returned to average. And when he decided to reach for the magazine for another look, the same reaction was repeated also in the cells.

A similar correlation was noted when a retired U.S. Navy gunner was tested who had been at Pearl Harbor during the Japanese attack. When watching a TV program titled *The World at War,* the gunner did not react to the downing of enemy aircraft by naval gunfire. He did react, however, when the downing occurred immediately following a facial close-up of an American naval gunner in action. At that point— when he seemed to have projected his own wartime experiences into the scene—his cells, located at a distance of eight miles, showed a reaction that was precisely correlated with his own.

As discussed in chapter 9, persistent nonlocal connection between cells removed from an organism and the host organism was discovered also in an independent set of experiments by the Psionic Medical Society in the United Kingdom (now called the Lawrence Society for Integral Medicine (Psionic Medical Society 2000). The qualified medical doctor members of the society sought access to what they call the "psi field," a nonlocal field they say envelops the organism. To access information in the psi field, the physicians make use of a sophisticated form of medical dowsing. They found that they could diagnose their patients at any distance; all they needed was a so-called "witness," which could be any protein sample from the body of the patient, such as a strand of hair or drop of blood. They could perform the diagnosis by observing the movement of a pendulum over a specially designed chart. In thousands of cases, decoding the pendulum's movement produced a correct diagnosis of the patient's condition.

The cells that make up the witness can be analyzed repeatedly, at any time and at any distance from the patient. The information they yield reflects the patient's state of health at the time that analysis is carried out, and not at the time the cells were removed from the body of the patient. This suggests that it is not the actual condition of the cells that conveys the information—because the information would then reflect the condition of the patient when the cells were removed. Instead they reflect the patient's physical condition at the time the tests are carried out. It appears that the cells remain nonlocally connected with the organism from which they have been removed.

These findings, though surprising on first sight, stand to reason. Superfast distance-independent connection among the parts of the organism is essential if the organism is to maintain the coherence it requires to sustain itself in its physically unstable living state far from thermal and chemical equilibrium. Narrow-band and relatively slow neural and biochemical signal-transmission cannot ensure by itself an adequate level of coherence in the organism. Only the nonlocal "entanglement" of the organism's cellular and subcellular components can create a sufficiently rapid flow of multidimensional information to maintain the living state.

Nonlocal Connection between Individuals

Nonlocal connections also surface between discrete organisms located at distances that preclude physical and physiological connection between them.

Experiments using functional magnetic resonance imaging (fMRI) tested the contact between the cerebral activity of individuals who were not in any known form of communication with each other. Jeanne Achterberg of the Saybrook Institute asked eleven healers to choose persons with whom they felt an empathic connection (Achterberg, Cooke, Richards, et al. 2005). The selected subjects were placed in an MRI scanner isolated from sensory contact with the healers. The healers sent energy, prayer, or good intentions—so-called distant intentionality—at random intervals that were unknown to both the healers and the subjects. In analyzing the results Achterberg found significant differences between the "send" and "no send" (control) periods in the activity of various parts of the subjects' brain, namely the anterior and middle cingulate areas, the precuneus, and the frontal areas. The probability that these differences would have occurred by chance was calculated as one in ten thousand.

Tests regarding the effects of conscious intention offer further evidence for nonlocal connection between individuals. It has been known for a long time that the focused intention of one person can affect the bodily state of another. This was confirmed by anthropologists researching what is known as "sympathetic magic" in traditional cultures. In his famous study *The Golden Bough,* Sir James Frazer noted that Native American shamans practicing voodoo would draw the figure of a person in sand, ashes, or clay, then prick it with a sharp stick or do it some other injury. The corresponding injury was said to be inflicted on the person the figure represented. Anthropologists found that the targeted person often fell ill, became lethargic, and would sometimes soon die. Dean Radin at the University of Nevada tested the positive variant of this effect under laboratory conditions.

In Radin's experiments the subjects created a small doll in their own image and provided various objects (pictures, jewelry, an autobiography, and personally meaningful tokens) to "represent" them. They also gave a list of what makes them feel nurtured and comfortable. The list and the information that accompanied it were used by the healers to create a sympathetic connection to the subjects, who were wired to monitor the activity of their nervous system—electrodermal activity, heart rate, and blood pulse volume—while the healer was in an acoustically and electromagnetically shielded room in an adjacent building. The healer placed the doll or the other objects provided by the subjects on a table and concentrated on them while sending randomly sequenced "nurturing" (active healing) and "rest" messages to the subjects.

In the experiments the electrodermal activity and heart rate of the subjects were significantly different during the active nurturing periods and during the rest periods, while blood pulse volume was significant for a few seconds during the nurturing period. Both heart rate and blood flow indicated a "relaxation response"—which made sense since the healer was attempting to "nurture" the subject via the doll. On the other hand, a higher rate of electrodermal activity showed that the subject's autonomic nervous system was becoming aroused. Why this should be so was puzzling until the experimenters realized that the healers nurtured the subjects by rubbing the shoulders of the dolls that represented them, or stroked their hair and face. For the subjects this worked like a "remote massage." Radin concluded that the actions of the healer reach the distant subject almost as if healer and subject were next to each other.

Radin's findings have been corroborated by William Braud and Marilyn Schlitz in hundreds of experiments carried out over more than a decade. The experiments tested the impact of the mental imagery of senders on the physiology of receivers. The effects proved similar to those produced by the subjects' own mental processes on their own body. "Telesomatic" action by a distant person proved to be nearly as effective as "psychosomatic" action by the subjects themselves.

Experiments that demonstrate the effectiveness of distant healing were pioneered by cardiologist Randolph Byrd (1988), a former professor at the University of California at Berkeley. His ten-month computer-assisted study concerned the effects of intention on patients admitted to the coronary care unit at San Francisco General Hospital. Byrd formed a group of experimenters made up of ordinary people whose only common characteristic was a habit of regular prayer in Catholic or Protestant congregations around the country. The selected people were asked to pray for the recovery of a group of 192 patients; another set of 210 patients for whom nobody prayed made up the control group. Strict criteria were used: the selection was performed double-blind: neither the doctors nor the nurses knew which patients belonged to which group. The experimenters were given the names of the patients, and some information about their heart condition and were asked to pray for them every day. They were not told anything further. Since each experimenter could pray for several patients, each patient had between five and seven people praying for him or her.

The results were statistically significant. The prayed-for group was five times less likely than the control group to require antibiotics (3 versus 16 patients); it was three times less likely to develop pulmonary edema (6 compared to 18 patients); none in the prayed-for group required endotracheal intubation (while 12 patients in the control group did); and fewer patients died in the former than in the latter group (though this particular result was statistically not significant). It did not matter how close or far the patients were to those who prayed for them, nor what type of praying was practiced—only the fact of concentrated and repeated prayer seemed to have counted, without regard to whom the prayer was addressed and where it took place.

In the form of alternative medicine that physician Larry Dossey calls "Era III nonlocal medicine," nonlocal effects are used systematically for healing (Dossey 1989). For example, a sensitive is asked to concentrate on a given patient from a remote location. As shown by the practice of various healers, it is enough to give the name and date of birth of the patient. Neurosurgeon Norman Shealy would often telephone this information from his office in Missouri to clairvoyant diagnostician Caroline Myss in New Hampshire. She diagnosed the cases and sent the results to Dr. Shealy. The latter found that in the first one hundred cases her diagnosis was ninety-three percent correct.

The above examples merely sample the variety of cases where nonlocal effects appear. They testify that nonlocality in the living world is not an exceptional phenomenon: it is a basic form of interaction, the essential precondition of the emergence and evolution of life in the biosphere.

Nonlocal Connections between Organisms and Environments

Nonlocal connections between organisms and environments are not recognized in mainstream biology: even the possibility of their existence is denied. Organisms are related to their environment, and to the world beyond the organism, through and only through interactions conveyed by the classical fields and biochemical processes. Yet it turns out that even photosynthesis, the basis of all life on the planet, relies on quantum processes. Gregory Engel and his collaborators (Engel, Calhoun,

Read, et al. 2007) found that long-lived electronic quantum coherence is required to explain the extreme efficiency of the photosynthetic process by allowing molecular complexes to sample vast areas of phase space and find the most efficient path.

More general evidence for the existence of nonlocal connections surfaces in regard to the course of biological evolution. In their absence Darwin's classical theory of the evolution of species through chance mutations in the genome cannot account for the record of phylogenetic evolution on Earth. The oldest rocks date from about four billion years before our time, while the earliest and already highly complex forms of life (blue-green algae and bacteria) are over 3.5 billion years old.

Half a billion years seems like a long time, but it does not appear to be long enough to explain how complex species could have evolved by chance mutations. The assembly even of a primitive prokaryote involves building a double helix of DNA consisting of some one hundred thousand nucleotides, with each nucleotide containing an exact arrangement of thirty to fifty atoms, together with a bilayered skin and the proteins that enable the cell to take in food. This construction requires an entire series of reactions, finely coordinated with each other. If living species had relied on random variation in an isolated genome, the level of complexity we observe in the domain of life is not likely to have been achieved in the approximately half a billion years that was available for it.

The negative probability that random chance events would produce complex species within realistic timeframes is heightened by additional considerations. It is not enough for mutations to produce one or a few positive changes in the genetic pool of a species; if the changes are to be viable, they must involve the entire genome. The evolution of feathers, for example, doesn't produce a reptile that can fly; radical changes in musculature and bone structure are also required, along with faster metabolism to power sustained flight. These changes involve complex processes: at least nine varieties of genetic rearrangements are known (transposition, gene duplication, exon shuffling, point mutation, chromosomal rearrangement, recombination, crossing over, pleiotropic mutation, and polyploidy). Many of them are interrelated. It is not very likely that these rearrangements, whether they occur singly or in combination, would produce new species from the old by chance variations of the genome. A random mutation is not likely to result in evolutionary advantage; it is likely to make the species less rather than more fit to survive. And in that case it would ultimately be eliminated by natural selection. Yet many species turned out to be viable, and their evolution was far more rapid than the vast search space of chance variation would allow.

The synthetic theory is further challenged by evidence that the genome is not fully isolated from the phenome, and hence it cannot mutate in isolation. But even the "fluid genome," responsive to influence from the phenome, fails to explain the precise, rapid, and highly focused macromutations called for to transform the genetic information of an unviable species into information that codes a new and viable species. If the genome-phenome system is to produce the necessary genetic mutations it must be sensitive to changes in the organism's milieu and produce adaptive responses to them. This sensitivity, however, is beyond the scope of the currently known interaction between organisms and environments.

The mainstream paradigm cannot explain how new and viable species could have emerged through random mutations in the genome. According to mathematical physicist Sir Fred Hoyle, the probability that this would be the case is about the same as the probability that a hurricane blowing through a scrapyard would assemble a working airplane.

As early as 1937, biologist Theodosius Dobzhansky noted that the birth of new species by genetic mutation would be impossible in reality, even if the birth of new species occurred on a "quasi-geological scale." However, new species appear much faster than that. In their theory of "punctuated equilibrium" Stephen Jay Gould and Niles Eldredge noted that the populations that are most likely to mutate are peripherally isolated and relatively small. Changes in their genome are fast and yet precise, often taking no more than five to ten thousand years. This makes the geological time of species evolution into an insignificantly short time—an evolutionary instant.

The evolution of species, the same as the self-maintenance of individual organisms, cannot rely on the known varieties of biochemical interaction alone. This is indirect but cogent evidence that the evolution of higher forms of life must involve quasi-instant multidimensional connection between organisms and the parts of organisms.

Nonlocal Connections and the Presence of Life

For most of the twentieth century, scientists believed that life in the universe is a chance occurrence, due to the lucky coincidence that a set of in-themselves-improbable conditions came together. It was known that for life to appear not only did the basic constants and parameters of the universe need to be finely adjusted, but further conditions also needed to be met. There must be a planet with the correct mass at the right distance from a main sequence G2 dwarf star; the planet needs to occupy a nearly circular orbit; it must have an oxygen/nitrogen rich atmosphere, a large moon, and a moderate rate of rotation; it must be at the right distance from the center of the galaxy and have liquid water on its surface; and it must have a correct ratio between water and landmass. Last but not least, it must have protection from asteroids through giant gas planets in its solar system.

It was held that the requirement for such improbably coinciding factors makes life an exceptional event in the universe. But this view was placed in question by a surprising finding published in October of 2011. A team of researchers headed by Sun Kwok and Yong Zhang (2011) at the University of Hong Kong reported that organic molecules, the basic building blocks of life, are created in stars. Some 130 such molecules have been found at the time of writing, including glycine, which is an amino acid, and ethylene glycol, the compound associated with the formation of the sugar molecules necessary for life.

Working with NASA's Spitzer Space Telescope, the astrophysicists found that water, methanol, and carbon dioxide coat dust particles around stars in the constellation Taurus, four hundred light years from Earth. These substances show up in interstellar dust clouds and dusty planet-forming discs around stars. It appears that at various stages of their evolution active stars eject organic compounds into interstellar space, distributing the molecules over vast regions. NASA noted that, while such materials have been found elsewhere, "this is the first time they were seen unambiguously in the dust making up planet-forming gases." This observation is an anomaly not only for the standard theory of cosmology, but for contemporary astrophysics as a whole. Yet, as Kwok et al. write, "Our work has shown that stars have no problem making complex organic compounds under near-vacuum conditions." Whereas this is theoretically impossible, observationally it is shown to be happening.

The observation that organic molecules are synthesized in stars is a discovery of major significance. It tells us that life is not an anomaly in the universe: the very processes that underlie the evolution of stars provide the template for the evolution of biological systems. It appears that only the higher forms of life—complex biochemical systems capable of metabolism and reproduction—require the kind of highly specific combination of conditions that are likely to be rare in the universe.

Nonlocal Connections and Mainstream Biology

Quantum-type connections within and between organisms are not recognized in mainstream biology. In regard to processes within the organism they are not considered necessary, and with respect to connections among distant individuals mainstream biologists consider the evidence spurious.

Local and deterministic interaction within the organism is said to be sufficient to account for the facts. One part of the organism is believed to determine the state of the other parts. This, however, is often not the case. Molecular interactions do not rigorously determine functions and processes in the organism. Even the genetic variety of determinism (the doctrine that genes in the organism contain the full set of instructions for building the whole organism) is inadequate to explain the observed facts. While it is true that genes determine the amino acid sequence of protein molecules by means of creating copies of messenger RNAs, this does not fully determine the way the organism functions. The organism is a highly integrated system where a score of processes engage all levels simultaneously, from the microscopic to the molecular and the macroscopic. Adjustments, responses, and changes required for the maintenance of the system propagate in all directions simultaneously and are sensitively tuned to conditions in the organism's environment. Some basic developmental processes are either entirely outside genetic control or are only indirectly affected by genes. In a collection of studies published by Hans-Peter Dürr, Fritz-Albert Popp, and Wolfram Schommers, Russian biophysicist Lev Beloussov suggested that the truth may be the reverse of genetic determinism: genes themselves could be obedient servants fulfilling powerful commands by the rest of the organism (Beloussov 2002).

Genetic determinism faces the C-value paradox (where C stands for complexity and C-value denotes the size of the organism's haploid set of chromosomes; that is, the size of its DNA sequence), as well as the gene-number paradox (the paradox of gene redundancy). In regard to C-value, the empirical findings are contrary to expectations. If the information coded in the genome would provide a more-or-less complete description of the organism, the complexity of the phenome (the flesh-and-blood organism) and the complexity of the genome should be proportional: more complex organisms should have more complex genetic information. But the complexity of the genome and the complexity of the organism are not correlated. The Human Genome Project identified less than forty thousand genes in the human genome, a surprisingly small amount. Even an amoeba has two hundred times more DNA per cell than a human being.

It is also puzzling that phylogenetically closely related species should have radically different genomes. The size of the genome in highly related rodents often varies by a factor of two, and the genome of the housefly is five times larger than the genome of the fruit fly. At the same time some phylogenetically distant organisms have similar genetic structure. Given these discrepancies, it is difficult to see how the structure of the genome would determine the structure of the phenome. One and the same gene can produce different proteins, and different genes can produce one and the same protein. The complex and coherent functioning of the living organism cannot be explained uniquely, or even primarily, in reference to genetic instruction governing molecular interaction.

Recent observations show that the living state is maintained not by genetic, but primarily by *epigenetic* regulation. Epigenetic regulation does not change the sequence of genes in the DNA but determines the effect of the sequence on the organism. In this way even genetic information is alterable in and by the organism. And inasmuch as the organism is in continuous communication with its environment, genetic information is also alterable by the organism's environment. Epigenetic regulation turns the genes in the organism "on" or "off" according to need. There is observational evidence that this kind of regulation can be handed down to succeeding generations.

Further observations indicate that in the organism the channel of communication, the vehicle that ensures its macroscopic coherence, is water. The human organism is more than seventy percent water, and this water is not the same liquid as water in the environment: it is both structurally and dynamically different (Del Giudice and Pulselli 2010). The ensemble of the electromagnetic waves released in the organism during a biochemical cycle is imprinted in the water within it. The organism "informs" the water it contains, and the informed water creates channels of communication in the whole organism. The transfer of information between informed water in the organism and its cells and cellular systems harmonizes the phase of the waves emitted by the latter, and this contributes to the organism's macroscopic coherence.

This water-transmitted harmonization is a nonlocal process: it does not involve any known form of energy.

The findings that come to light regarding the biophysics of water, together with the puzzles that confront molecular and genetic determinism, corroborate Hans Fröhlich's bold hypothesis: all parts of the living system create fields at various frequencies, and these propagate throughout the organism. Through informed water, the specific resonance frequency of molecules and cells in the organism is harmonized, and long-range phase correlations come about. These are similar to, although not as pronounced as, the correlations that come about in superfluidity and superconductivity.

THE NONLOCALLY INTERCONNECTED UNIVERSE

We conclude this concise review of the evidence for nonlocality and interconnection in the principal fields of scientific inquiry with a large-scale picture of the world, a picture in which nonlocality and interconnection are universal phenomena. We begin this review with an outline of the so-called standard model of the universe.

The Standard Model

According to the standard model in physical cosmology, a single event created the world, a nonrecurring and unexplained singularity known as the Big Bang. The universe is said to have originated in that explosive instability 13.75 ± 0.13 billion years ago. A region of cosmic prespace had exploded, creating a fireball of staggering heat and density. In the first few milliseconds it synthesized all the matter that populates space and time. The particle-antiparticle pairs that emerged from the cosmic prespace collided with and annihilated each other; the one billionth of the originally created particles that survived the collisions (the tiny excess of particles over antiparticles) constitutes what we call matter in the universe.

After about four hundred thousand years photons decoupled from the radiation field of the primordial fireball: space became transparent, and clumps of matter established themselves as distinct entities. Due to gravitational attraction these clumps condensed into stars and stellar systems and created gigantic swirls that, after about a billion years, became galaxies.

Anomalies in the Standard Model

The universe that evolved in the wake of the primal explosion turned out to be precisely balanced between contraction and expansion. The computer analysis of some three hundred million observations first made through NASA's Cosmic Background Explorer Satellite (COBE) in 1991 provided remarkable confirmation. Detailed measurements of the cosmic microwave background radiation—the presumed remnant of the Big Bang—indicate that the variations are not distortions caused by radiation from stellar bodies but are the remnants of minute fluctuations in the cosmic fireball when it was less than one trillionth of a second old. In April 1992, a team of astrophysicists led by George Smoot (Smoot and Davidson 1994) announced that they had mapped the intensity of the radiation from the primal fireball and found tiny variations that, under the action of gravitation, acted as the seeds from which the macrostructures of the universe had evolved.

Measurements of the background radiation disclose the matter-density of the universe; that is, the number of particles that were created in the primal explosion and were not annihilated in the particle-antiparticle collisions that followed. If the surviving particles make for a matter-density above a certain number (estimated at 5×10^{-26} g/cm^3), the gravitational pull associated with the total amount of matter will ultimately exceed the inertial force generated by the primal explosion, and the universe is "closed": it will collapse back on itself. If matter-density is below that number, expansion will continue to dominate gravitation. Then the universe is "open": it will expand indefinitely. However, if matter-density is precisely at the critical value, the forces of expansion and contraction will balance each other and the universe is "flat": it will remain balanced at the razor's edge between the opposing forces of expansion and contraction.

Recent findings show that the universe created in the wake of the primal explosion was fine-tuned to the staggering precision of one part in 10^{50}.

Sophisticated measurements further disclose that not only is the matter-density of the universe precisely tuned for balance between expansion and contraction; the existing particles are also finely tuned to the macroparameters of the universe. Applying Einstein's celebrated relation between mass and energy ($E = mc^2$), the size of the electron (which is $ro = 6.10^{-15}$ meters) turns out to be a consequence of the number of electrons in the universe (the latter is given by Eddington's number, approx. 2×10^{79} in the Hubble universe of $R = 10^{26}$ meters).

Moreover also the physical constants of the universe are surprisingly tuned to each other. Physicist Dezsö Sarkadi discovered that a simple exponential rule ($Q = 2/9$) is maintained when the relation of one constant to the other is quantified (Sarkadi and Bodonyi 1999).

Menas Kafatos and collaborators (Kafatos and Nadeau 1999) came up with further relations between the macroparameters of the universe. Scale invariant relations appear between the mass constituted by the total number of particles in the universe, the gravitational constant, the charge of the electron, Planck's constant, and the speed of light. Among other things, all lengths turn out to be proportional to the scale of the universe. This suggests a staggering level of coherence among the constants of the cosmos. Kafatos states unequivocally that the entire universe is instantly interconnected and hence nonlocal.

Unexpected forms of coherence, and by implication nonlocality, come to light also in the processes that lead to the evolution of life. The basic forces and constants of the universe are precisely tuned to create conditions that permit the kind of complex systems on which life is based to emerge. A minute difference in the strength of the electromagnetic field relative to the gravitational field would have prevented the evolution of such systems, since hot and stable stars such as our Sun could not have come about. If the difference between the mass of the neutron and the proton were not precisely twice the mass of the electron, no substantial chemical reactions could take place. And if the electric charge of electrons and protons did not balance precisely, all configurations of matter would be unstable and the universe would consist only of radiation and a relatively uniform mixture of gases.

In this universe the gravity constant ($G = 6.67 \times 10^{-11}\,\mathrm{Nm^2/s^2}$) is precisely such that stars can form and shine long enough to allow the evolution of the structure of galaxies, including solar systems within the galaxies. If G were smaller, particles would not compress sufficiently to achieve the temperature and the density needed to ignite hydrogen: stars like our Sun would have remained in a gaseous state. If on the other hand G were larger, stars would have formed but would burn faster and endure for a shorter time, making it unlikely that life could evolve on the planets orbiting them.

Likewise, if the Planck constant ($h = 6.62606957 \times 10^{-34}$ sec) were even minutely other than it is, carbon-producing nuclear reactions could not occur in stars—and consequently complex systems based on carbon-bonding could not arise on some of their planets. But given the actual value of G and h, and of an entire array of universal constants (including the velocity of light, the size and mass of the electron, and the relationships between the size of the proton and the nucleus), complex systems and thus life could in fact arise on some planets orbiting active stars. This level of coherence, and thus the very presence of life, are beyond the scope of the standard model: they are an unexplained anomaly.

A further feature of coherence in the universe is the uniformity of the background radiation. The microwave background radiation, emitted when the universe was about a hundred thousand years old, is known to be isotropic, the same in all directions. This calls for an explanation. Physical signals could not have tuned the various regions of the expanding universe together, because when the background radiation was emitted the universe's opposite sides were already ten million light-years apart—and light could only have traveled one hundred thousand light-years in that time. Although they were not connected by classical forces and signals beyond the first few microseconds after the explosion that created our universe, even distant galaxies and other macrostructures evolve in a uniform manner.

Cosmologist Alan Guth (1997) proposed the (rather mind-boggling) theory of inflation to account for this "horizon problem." According to this theory, in the span of the 10^{-33} sec Planck time following the birth of the universe, space expanded at a rate faster than light. This did not violate general relativity, since it was not matter that moved at a supraluminal velocity but space—matter stood still relative to space. During the initial Planck time all parts of the universe were in contact, sharing the same density and temperature. Then, as the universe expanded, some parts moved beyond the range of physical contact and evolved independently. But they could evolve uniformly, since they were connected at the birth of the universe.

Multiverse Models

Cosmologies that conceive of the universe as a cycle in a vaster and conceivably infinite "multiverse" have appeared in the past few decades, and in the opinion of many cosmologists they are an improvement over the standard single-universe model.

The term *multiverse* was first used by William James in 1895 (in *The Will to Believe and Other Essays in Popular Philosophy*), but the idea itself had been advanced previously by Nicolas of Cusa in the fifteenth century. It was taken up by Giordano Bruno in his 1584 treatise *On the Infinite Universe and Worlds*. Bruno pointed out that if we consider the universe a single infinity we must necessarily distinguish between "universe" and "world." The "universe," he wrote, is a single infinity that harbors a plurality of "worlds."

One of the first modern multiverse models was the work of physicist John Wheeler. In his model, advanced more than three decades ago, the expansion of the universe comes to an end, and our universe col-lapses back on itself. Following this "Big Crunch" a new expansionary phase gets under way. In the quantum uncertainties that dominate the supercrunched state nearly infinite possibilities exist for the emergence of a new universe.

New universes could also be created inside black holes. The extreme high densities of these spacetime regions represent recurrent singulari-ties where the laws of physics do not apply. Stephen Hawking and Alan Guth suggested that under these conditions the black hole's region of spacetime detaches itself from the rest and expands to create a universe of its own. Thus the black hole of one universe may be the "white hole" of another: it is the "Bang" that creates it.

Still another multiverse cosmology was elaborated by Ilya Prigogine, J. Geheniau, E. Gunzig, and P. Nardone (1988). Their theory suggests that major matter-creating bursts occur from time to time. The large-scale geometry of spacetime creates a reservoir of "negative energy" (which is the energy required to lift a body away from the direction of its gravitational pull), and from this reservoir gravitating matter extracts positive energy. Thus gravitation is at the root of the ongoing synthe-sis of matter: it produces a perpetual matter-creating mill. The more particles are generated, the more negative energy is produced and then transferred as positive energy to the synthesis of still more particles.

Given that the cosmic matrix ("vacuum") is unstable in the presence of gravitational interaction, matter and matrix form a self-generating feedback loop. A critical matter-triggered instability causes the matrix to transit to the inflationary mode, and that mode marks the beginning of a new era of matter synthesis—a new universe.

Other cosmologies postulate the continuous creation of matter, among them the QSSC (Quasi-Steady State Cosmology) advanced by Fred Hoyle with Geoffrey Burbidge and J. V. Narlikar (1993). Hoyle's original 1983 model called for a continuous, relatively linear creation of matter in space in which baby universes are periodically created in bursts similar to that which brought forth our own universe. These "matter-creating events" are interspersed throughout the universe.

As the original version of the theory encountered empirical discrepancies, it was replaced by a version that called for matter creation preferentially in regions of high matter density where gravity is strongly negative. The revised theory suggests that the most recent burst of matter creation occurred about fourteen billion years ago, in good agreement with independent estimates of the age of our universe.

A more recent continuous creation theory emerges from subquantum kinetics, a reaction-diffusion ether theory proposed by Paul LaViolette (see Appendix II, below). As in the QSSC, subquantum kinetics requires that matter is created most rapidly in regions of high matter-density, although this density dependent aspect emerges as a prediction of the physics of the theory rather than as an added assumption. Whereas QSSC assumes that continuous creation proceeds within a continuously expanding space, subquantum kinetics assumes that the universe is not expanding. It interprets the cosmological redshift, not as a recessional Doppler effect, but as a tired-light energy loss effect, based on the theory's fundamental equation system.

A particularly sophisticated multiverse cosmology has been advanced by Paul J. Steinhardt of Princeton and Neil Turok of Cambridge (Steinhardt and Turok 2002). Their model accounts for all the facts accounted for by the standard model, and it also explains an observation that is an anomaly for the latter: the accelerating expansion of distant galaxies. According to Steinhardt and Turok, the universe undergoes an endless sequence of cosmic epochs, with each epoch beginning with a "Bang" and ending with a "Crunch." Each cycle goes through a period of gradual and then accelerating expansion, followed by reversal and contraction.

Steinhardt and Turok estimate that at present we are about fourteen billion years into the current cycle, at the beginning of a trillion-year period of continuous, and continuously accelerating, expansion. Ultimately space becomes homogeneous and flat, and a new cycle gets under way.

Cosmologist Leonard Susskind (2006) suggested that the staggering number of universes implied by M-theory, a recent version of string theory, is not a flaw in that theory but a profound insight into the nature of reality: each solution of M-theory's equations corresponds to a real universe, with all its laws and constants. The full range of universes governed by all possible laws is the "landscape," and the collection of the universes described by these laws is the multiverse.

The same idea is at the core of Andrei Linde's (1990, 2004) version of inflation theory. According to Linde, the superrapid explosion that accompanied the birth of our universe was reticular, made up of several individual regions. What we know as the Big Bang had distinct regions, much like a soap bubble in which smaller bubbles are stuck together. As such a bubble is blown up, the smaller bubbles separate, each forming a distinct bubble of its own. The bubble universes percolate outward and follow their own evolutionary path.

Each universe produces its own physical constants, and these may be very different from one another. For example, in some universes gravity may be so strong that material structures recollapse almost instantly; in others, gravity may be so weak that no stars will form. Our bubble universe happens to provide just the right conditions for the evolution of complex systems, and hence of life. The basic idea of this theory received a boost in 2011, when Hiranya Peiris, a cosmologist at London's University College, worked out that the creation of bubble universes would leave characteristic patterns in the cosmic microwave background (CMB), and these could be detected by the Planck telescope. Patterns that may be signatures of bubble universes have then been found in the CMB, although their full verification is yet to be achieved.

An analogous idea underlies the cosmologies advanced by Lee Smolin, Stephen Hawking, Steven Weinberg, and Max Tegmark, among others. According to Martin Rees, the current "multiverse revolution" is just as profound as the Copernican Revolution was in the seventeenth century.

The Akasha Paradigm in Physics

Two Hypotheses

According to the Akasha paradigm, the observable phenomena of the world are manifestations of an in-itself-unobservable fundamental dimension. It is in that dimension that manifest phenomena arise, and it is in that dimension that the laws of their interaction are encoded.

Demonstrating the validity of these claims has been beyond the scope of physics until recently. With the emergence of the holographic spacetime theory and the parallel discovery of the amplituhedron (see chapter 4), the hypothesis of a primal dimension beyond spacetime has become part of the discourse of cutting-edge physics, in particular quantum field physics. The insight that dawns is that the events we observe in spacetime are coded beyond spacetime. According to the interpretation offered here, they are coded in the Akasha: the "deep" or "hidden" dimension of the cosmos.

The first hypothesis, by Paul LaViolette, addresses the dynamics whereby the phenomena we encounter in spacetime are generated in the deep domain he calls "transmuting ether." His subquantum kinetics postulates the presence of intrinsically unobservable subquantum entities called "etherons." In their interaction etherons create the physical constants of the universe.

The second hypothesis, by Peter Jakubowski, offers a further demonstration that the whole of contemporary physics, with all its equations, units, and constants, can be derived from the Akashic deep dimension *alias* Universal Quantum Field. In their different ways and with their diverse methods, the two hypotheses make the same point: the phenomena observed in spacetime originate in, and need to be referred to, processes and relations that are beyond spacetime, in a dimension we name in honor of a classical insight *the Akasha*.

HYPOTHESIS 1:
THE TRANSMUTING ETHER
Paul A. LaViolette

The Akasha dimension is the "transmuting ether," an active cosmic matrix that gives rise to physical form. Its multifarious components, called "etherons," react among themselves, transforming and diffusing throughout space. Their interweaving processes bind the ether into organic unity.

Etherons in themselves have no mass, charge, or spin. The properties mass and spin appear when the etherons self-organize into soliton fields, etheric concentration patterns, which we recognize as material particles. Neutrons spontaneously nucleate from large magnitude electric and gravity fluctuations in the ether's vacuum state. Charge arises subsequently as a secondary organizational restructuring when a neutron progenitor spontaneously transforms into a positively charged proton with the emission of a negatively charged electron and antineutrino.

The ether, as the Akasha dimension, is the ultimate reality; its concentration gradients are the prime cause of motion. Force is a derived stress effect (a distortion of the subatomic particle's etheric soliton pattern) engendered when a concentration gradient is superimposed. —E.L.

THE NEW CONCEPT OF THE ETHER

Subquantum kinetics is a unified field theory whose description of microphysical phenomena has a general systems theoretic foundation (LaViolette 1985a, 1985b, 1985c, 1994, 2013). It conceives subatomic particles to be localized Turing wave patterns that self-organize within a subquantum medium that functions as an open reaction-diffusion system. This nonexpanding medium, termed the *transmuting ether* forms the substrate from which all physical form in our universe emerges. This substrate, which requires more than three dimensions for its description, differs from nineteenth century mechanical ethers in that it is continually active, its multifarious components transmuting, reacting among themselves, and diffusing through space, these interweaving processes binding the ether into an organic unity.

Subquantum kinetics presents a substantially different paradigm from that of standard physics, which views particles as closed systems. Whether these be subatomic particles bound together by force fields, or quarks bound together by gluons, physics has traditionally conceived nature at its most basic level to be composed of immutable structures. Unlike living systems, which require a continuous flux of energy and matter with their environment to sustain their forms, conventional physics has viewed particles as self-sufficient entities that require no interaction with their environment in order to continue their existence. Thus classical field theory leads to a conception of space that Alfred North Whitehead has criticized as being one of mere simple location, where objects simply have position without incorporating any reference to other regions of space and other durations of time.

Whitehead instead advocated a conception of space manifesting prehensive unification, where separate objects can be "together in space and together in time even if they be not contemporaneous." The ether (as the Akasha) of subquantum kinetics fulfills Whitehead's conception. As shown below, it is precisely because of its nonlinear, reactive, and interactive aspect that the transmuting ether of subquantum kinetics is able to spawn subatomic particles and photons, manifested either as stationary or inherently propagating ether concentration patterns. In the context of subquantum kinetics, the very existence of the physical world we see around us is evidence of the dynamic organic unity that operates in the universal substratum below, imperceptible to us and out of reach of direct detection by the most sophisticated instruments. The Akasha concept presented in this book embraces Whitehead's organic conception as well as the reaction-diffusion ether concept embodied in subquantum kinetics and lays the basis for an exciting new unified science paradigm.

The notion of an ether, or of an absolute reference frame in space, necessarily conflicts with the postulate of special relativity that all frames should be relative and that the velocity of light should be a universal constant. However, experiments by Sagnac (1913), Graneau (1983), Silvertooth (1987, 1989), Pappas and Vaughan (1990), Lafforgue (1991), and Cornille (1998), to name just a few, have established that the idea of relative frames is untenable and should be replaced with the notion of an absolute ether frame. Also a moderately simple experiment performed by Alexis Guy Obolensky has clocked speeds as high as 5 c (c = speed of light) for Coulomb shocks traveling across his laboratory (LaViolette 2008a). Furthermore Podkletnov and Modanese (2011) report having measured a speed of 64 c for a collimated gravity impulse wave produced by a high voltage discharge emitted from a superconducting anode. These experiments not only soundly refute the special theory of relativity, but also indicate that information can be communicated at superluminal speeds.

However, subquantum kinetics does not negate the existence of "special relativistic effects" such as velocity dependent clock retardation and rod contraction. Nor, in offering an alternative to the spacetime warping concept of general relativity, does it negate the reality of orbital precession, the bending of starlight, gravitational time dilation, and gravitational redshifting. These effects emerge as corollaries of its reaction-diffusion ether model (LaViolette 1985b, 1994, 2004, 2013).

THE SYSTEMS DYNAMICS OF SUBQUANTUM KINETICS

Subquantum kinetics was inspired by research done on open chemical reaction systems such as the Belousov-Zhabotinskii (B-Z) reaction (Zaikin and Zhabotinskii 1970, Winfree 1974) and the Brusselator (Lefever 1968; Glansdorff and Prigogine 1971; Prigogine, Nicolis, and Babloyantz 1972; Nicolis and Prigogine 1977). Under the right conditions, the concentrations of the variable reactants of the Brusselator reaction system can spontaneously self-organize into a stationary reaction-diffusion wave pattern such as that shown in figure 1.1 on page 146. These have been called

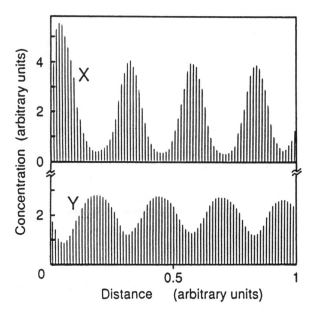

Figure 1.1. One-dimensional computer simulation of the concentrations of the Brusselator's X and Y variables (diagram after R. Lefever 1968).

Turing patterns in recognition of Alan Turing, who in 1952 was the first to point out their importance for biological morphogenesis. Alternatively, Prigogine et al. (1972) have referred to them as dissipative structures because the initial growth and subsequent maintenance of these patterns is due to the activity of the underlying energy-dissipating reaction processes. In addition, the B-Z reaction is found to exhibit propagating chemical concentration fronts, or chemical waves, which may be easily reproduced in a school chemistry laboratory; see figure 1.2.

The Brusselator, the simpler of the two reaction systems, is defined by the following four kinetic equations:

$$(1\text{-}a) \qquad A \xrightarrow{\ k_1\ } X$$

$$(1\text{-}b) \qquad B + X \xrightarrow{\ k_2\ } Y + Z$$

$$(1\text{-}c) \qquad 2X + Y \xrightarrow{\ k_3\ } 3X$$

$$(1\text{-}d) \qquad X \xrightarrow{\ k_4\ } \Omega$$

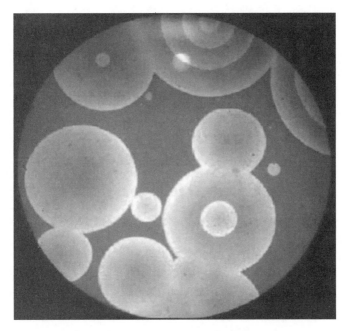

Figure 1.2. Chemical waves formed by the Belousov-Zhabotinskii reaction (photo courtesy of A. Winfree).

The capital letters specify the concentrations of the various reaction species, and the k_i denote the kinetic constants for each reaction. Each reaction produces its products on the right at a rate equal to the product of the reactant concentrations on the left times its kinetic constant. Reaction species X and Y are allowed to vary in space and time, while A, B, Z, and Ω are held constant.

This system defines two global reaction pathways that cross-couple to produce an X-Y reaction loop. One of the cross-coupling reactions, (1-c), is autocatalytic and prone to produce a nonlinear increase of X, which is kept in check by its complementary coupling reaction (1-b). Computer simulations of this system have shown that, when the reaction system operates in its supercritical mode, an initially homogeneous distribution of X and Y can self-organize into a wave pattern of well-defined wavelength in which X and Y vary reciprocally with respect to one another, as shown in figure 1.1. In other words, these systems allow order to spontaneously emerge (entropy to decrease) by virtue of the fact that they function as open systems, the second law of thermodynamics only applying to closed systems.

THE MODEL G ETHER
REACTION SYSTEM

Subquantum kinetics postulates a nonlinear reaction system similar to the Brusselator that involves the following five kinetic equations termed Model G (LaViolette, 1985b):

$$(2\text{-a}) \qquad A \underset{k_{-1}}{\overset{k_1}{\rightleftharpoons}} G$$

$$(2\text{-b}) \qquad G \underset{k_{-2}}{\overset{k_1}{\rightleftharpoons}} X$$

$$(2\text{-c}) \qquad B + X \underset{k_{-3}}{\overset{k_1}{\rightleftharpoons}} Y + Z$$

$$(2\text{-d}) \qquad 2X + Y \underset{k_{-4}}{\overset{k_1}{\rightleftharpoons}} 3X$$

$$(2\text{-e}) \qquad X \underset{k_{-5}}{\overset{k_1}{\rightleftharpoons}} \Omega$$

The kinetic constants k_i denote the relative propensity for the reaction to proceed forward, and k_{-i} denote the relative propensity for the corresponding reaction to proceed in the reverse direction. The forwarded reactions are mapped out in figure 1.3.

Since the forward kinetic constants have values much greater than the reverse kinetic constants, the reactions have the overall tendency to proceed irreversibly to the right. Nevertheless, the reverse reactions, in particular that associated with reaction (2-b), play an important role. Not only does this allow Model G to establish electrogravitic field coupling, but as described below, it also allows the spontaneous formation of material particles (i.e., dissipative solitons) in an initially subcritical ether.

Whereas the Brusselator and B-Z reaction conceive of a chemical

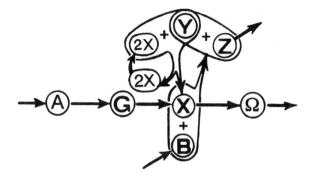

Figure 1.3. The Model G ether reaction system
investigated by subquantum kinetics

medium consisting of various reacting and diffusing molecular species, subquantum kinetics conceives of a space-filling etheric medium consisting of various reacting and diffusing etheric species termed *etherons*. Being present as various etheron types (or states) labeled A, B, X, and so on, these diffuse through space and react with one another in the manner specified by Model G. Model G is in effect the recipe, or software, that generates the physical universe, what Ervin Laszlo terms the "matrix" that generates the Manifest Dimension. Etherons should not be confused with quarks. Whereas quark theory proposes that quarks exist only within the nucleon, with just three residing within each such particle, subquantum kinetics presumes that etherons are far more ubiquitous, residing not only within the nucleon, but also filling all space with a number density of over 10^{25} per cubic fermi, where they serve as the substrate for all particles and fields.

The self-closing character of the X-Y reaction loop, which is readily evident in figure 1.3, is what allows Model G and the Brusselator to generate ordered wave patterns. Model G is similar to the Brusselator with the exception that a third intermediary variable, G, is added, with the result that steps (2-a) and (2-b) now replace step (1-a) of the Brusselator, all other steps remaining the same. The third variable was introduced in order to give the system the ability to nucleate self-stabilizing localized

Turing patterns within a prevailing subcritical environment.

This autogenic particle formation ability is what allows Model G to become a promising candidate system for the generation of physically realistic subatomic structures.

Based on the reaction equation system, we may write the following set of partial differential equations to depict how all three reaction intermediates G, X, and Y vary as a function of space and time in three dimensions, where the D_g, D_x, and D_y values represent the diffusion coefficients of the respective reactant variables.

$$(3\text{-}a) \qquad \frac{\partial G(x, y, z, t)}{\partial t} = k_1 A - k_2 G + D_g \nabla^2 G$$

$$(3\text{-}b) \qquad \frac{\partial X(x, y, z, t)}{\partial t} = k_2 G + k_4 X^2 Y - k_3 B X - k_5 X + D_x \nabla^2 X$$

$$(3\text{-}c) \qquad \frac{\partial Y(x, y, z, t)}{\partial t} = k_3 B X - k_4 X^2 Y + D_y \nabla^2 Y$$

These three relatively simple differential equations are the mathematical script that specifies that portion of the Akasha whose metabolic functioning forms the substrate for the emergence of a non-expanding physical universe. A homogeneous distribution of the G, X, and Y reaction intermediates would correspond to a spatial vacuum devoid of matter and energy. Variations in the concentrations of these three variables would correspond to the formation of observable electric and gravitational potential fields, and wave patterns formed by these fields, in turn, would constitute observable material particles and energy waves. The etherons themselves would remain unobservable. Subquantum kinetics identifies the G concentration with gravitational potential, where G concentrations greater than the prevailing homogeneous steady-state concentration value, G_o, would constitute positive gravity potentials and G concentrations less than G_o would constitute negative gravity potentials. A negative G potential well, i.e., G ether concentration well, would correspond to a matter-attracting gravity potential field, whereas a positive G potential hill would correspond to a matter-repelling gravity potential field.

The X and Y concentrations, which mutually interrelate in reciprocal fashion, are together identified with electric potential fields. A configuration in which the Y concentration is greater than Y_o and the X concentration is less than X_o would correspond to a positive electric potential, and the opposite polarity, low-Y/high-X, would correspond to a negative electric potential. Relative motion of an electric potential field, or of an X-Y concentration gradient, would generate a magnetic (or electrodynamic) force (LaViolette 1994, 2013). As Feynman, Leighton, and Sands (1964) have shown, in standard physics magnetic force can be mathematically expressed solely in terms of the effect that a moving electric potential field produces on a charged particle, obviating the need for magnetic potential field terms. Also relative motion of a gravity potential field, of a G concentration gradient, is predicted to generate a gravitodynamic force, the gravitational equivalent of a magnetic force.

The subquantum kinetics ether functions as an open system, where etherons transform irreversibly through a series of "upstream" states, including states A and B, eventually occupying states G, X, and Y, and subsequently transforming into the D and Ω states and from there through a sequence of "downstream" states (see figure 1.4). This irreversible sequential transformation is conceived as defining a vectorial dimension line termed the *transformation dimension*. Our observable physical universe would be entirely encompassed by the G, X, and Y ether states, which would reside at a nexus along this transformation dimension, the continual etheron transformation process serving as the Prime Mover of our universe.

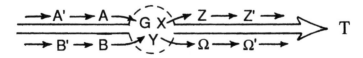

Figure 1.4. An expansion of the Model G ether reaction scheme as it would appear disposed along dimension **T**. Ether states G, X, and Y mark the domain of the physical universe.

According to subquantum kinetics, the arrow of time, as physically observed in all temporal events, may be attributed to the continuation of this subquantum transformative process. Subquantum kinetics allows the possibility of parallel universes forming either "upstream" or "downstream" of our own universe wherever the ether reaction stream intersects itself to form a reaction loop similar to Model G. However, while there is a finite chance of such a material universe being spawned, the possibility that it would actually form is vanishingly small since the ether reaction parameters would need to adopt the proper precise values in order to spawn the necessary nucleon building blocks.

Since etherons both enter and leave the etheron states that compose material bodies and energy waves in our physical universe, our observable universe may be said to be open to the through-put of etherons. That is, our universe would function as an open system. In such a system, ordered field patterns may spontaneously emerge from initially homogeneous field distributions or they may progressively dissolve back to the homogeneous state, depending on the criticality of the reaction system.

In Model G, the system's criticality is determined by the value of the G variable. Sufficiently negative G potentials create supercritical conditions that allow matter formation and photon blueshifting while positive G potential values that would be most prevalent in intergalactic space create subcritical conditions that cause tired-light photon redshifting which would account for the observed cosmological redshift.

The transmuting ether of subquantum kinetics bears some resemblance to the ether concept of Nikola Tesla. He proposed a gaslike ether that is acted on by a "life-giving creative force," which, when thrown into infinitesimal whorls, gives rise to ponderable matter; he also proposed that when this force subsides and the motion ceases, matter disappears, leaving only the ether. In subquantum kinetics, this creative force or Prime Mover is termed *etheric force,* while the resulting transmutative or reactive transformation of etherons from one state to another is termed *etheric flux.*

The transmuting ether also closely parallels the descriptions of

Besant and Leadbeater (1919), who as early as 1895 said "the ether is not homogeneous but consists of particles of numerous kinds, differing in the aggregations of the minute bodies composing them." About the subatomic particle, which they refer to as the "ultimate physical atom," they state: "It is formed by the flow of the life-force and vanishes with its ebb. When this force arises in 'space' . . . atoms appear; if this be artificially stopped for a single atom, the atom disappears; there is nothing left. Presumably, were that flow checked but for an instant, the whole physical world would vanish, as a cloud melts away in the empyrean. It is only the persistence of that flow which maintains the physical basis of the universe." Similarly, subquantum kinetics views our observable physical universe as an epiphenomenal watermark generated by the activity of a higher dimensional ether that functions as an open system.

PARTHENOGENESIS: THE CREATION OF MATTER FROM ZERO-POINT FLUCTUATIONS

According to subquantum kinetics, material particles nucleate from electric and gravity potential fluctuations that spontaneously arise from the ether vacuum state. Since etherons react and transform in a stochastic Markovian fashion, the etheron concentrations of all etheron species will vary stochastically above and below their steady-state values, the magnitudes of these fluctuations conforming to a Poisson distribution. It is known that such fluctuations are present in the chemical species of reaction-diffusion systems such as the B-Z reaction, and their presence is also postulated in the theoretical Brusselator system. So the same would be true in the Model G reactive ether. Hence subquantum kinetics predicts that stochastic electric and gravity potential fluctuations should spontaneously arise throughout all of space, in regions both where field gradients are present and where they are absent.

This is in some ways analogous to the concept of the zero-point energy (ZPE) background, but with some differences. In conventional physics, ZPE fluctuations are theorized to have energies comparable to the rest mass energy of subatomic particles and to emerge as particle-antiparticle pairs, which rapidly annihilate one another. As a result, it is fashionable to quote unimaginably high values of the order of 10^{36} to 10^{113} ergs/cm^3 for the zero-point energy density of space. But because of their polarity pairing, they are unable to nucleate matter. By comparison, subquantum kinetics rejects the idea that the spatial vacuum is "seething with virtual particles and antiparticles." It theorizes far lower ZPE densities, on the order of less than 1 erg/cm^3, or less than the radiation energy density prevailing at 2000 K. Nevertheless, because these fluctuations are unpaired, they are potentially able to spawn material particles. But this only occurs when a fluctuation of sufficiently large magnitude arises, the vast majority being far too small to attain the required subquantum energy threshold.

Provided that the kinetic constants and diffusion coefficients of the ether reactions are properly specified to render the system subcritical but close to the critical threshold, a sufficiently large spontaneously arising positive zero-point electric potential fluctuation (i.e., a critical fluctuation consisting of a low X concentration or a high Y concentration), with further growth, a further reduction of X and increase in Y, is able to break the symmetry of the initial vacuum state to produce what is called a Turing bifurcation; that is, it is able to change the initially uniform electric and gravity potential background field that defines the vacuum state into a localized, steady-state periodic structure. In subquantum kinetics this emergent wave pattern would form the central electric and gravity field structure of a nascent subatomic particle.

One advantage of Model G is that a positive electric potential fluctuation characterized by a negative X potential also generates a corresponding negative G potential fluctuation by virtue of the reverse reaction X $\xleftarrow{\quad k_{-2} \quad}$ G, and this in turn produces a local supercritical region allowing the seed fluctuation to persist and grow in size. Consequently, if the ether reaction system is initially in the subcritical vacuum state, provided that it operates sufficiently close to the critical threshold, eventually a fluctuation will arise that is sufficiently large to form a supercritical region and nucleate a subatomic particle (e.g., a neutron). Thus spontaneous matter and energy creation is allowed in subquantum kinetics.

Once formed, each proto neutron would undergo beta decay into a proton and electron which eventually would combine to form a hydrogen atom. Proto neutrons are predicted to have a greater probability of nucleating in the vicinity of an existing subatomic particle (e.g., proton) since its gravity potential well produces a fertile supercritical region. Hence hydrogen will tend to beget more hydrogen, and at times will transform into a higher mass nucleus to form also deuterium and helium nuclei. Unlike the Big Bang theory, primordial space is relatively cold, heated only by the beta decay energy released from sporadically emerging nascent neutrons. Consequently, the gas in each locale will eventually condense into a primordial planetesimal. Since subquantum kinetics predicts that particle creation proceeds more rapidly in the vicinity of existing matter, each planetesimal should eventually grow into a planet and then into a mother star which later spawns daughter planets and stars. With further growth and proliferation, these stars congregate into a primordial star cluster, which eventually grows into a dwarf elliptical galaxy. Later, as its central supermassive mother star begins explosive activity, it gradually transforms into a spiral galaxy, and finally into a giant elliptical.

Subquantum kinetics is incompatible with the expanding universe hypothesis or with the hypothesis of a Big Bang. It requires that the transmuting ether is cosmologically stationary and that galaxies, excepting their peculiar motions, are at rest relative to their local ether frame, for any cosmological expansion would cause the ether's reactant concentrations to progressively decrease over time and its state of criticality to become drastically altered. Moreover a Big Bang creation event is precluded since the emergence of a ZPE fluctuation large enough to create all the matter and energy in the universe in a single event would be a virtual impossibility.

This "parthenogenic," order-through-fluctuation process that produces nascent neutrons is shown in figure 1.5, which presents successive frames from a 3D computer simulation of equation system (3) (Pulver and LaViolette 2013). Spherical symmetry was imposed as an arbitrary assumption to reduce the computing time necessary to carry out the simulation. The duration of the simulation consists of 100 arbitrary time units and the reaction volume measures 100 arbitrary spatial units, from −50 to +50, with one-fifth of the volume being displayed in the graph. Vacuum boundary conditions are assumed. These space and time units are dimensionless, meaning that the units of measure are not specified. To initiate the particle's nucleation, a negative subquantum X ether fluctuation $-\varphi_x(r)$ was introduced at spatial coordinate r = 0. The rise and fall of the fluctuation magnitude reaches its maximum value of −1 after 10 time units, or 10% of the way through the simulation, and diminishes back to zero magnitude (flat line) by 20 units of time, or 20% of the way through the simulation. The reaction system quickly generates a complementary positive Y potential fluctuation $+\varphi_y(r)$, which together with the X fluctuation composes a positive electric potential fluctuation, and also generates a negative G fluctuation $-\varphi_g(r)$, which composes a gravity potential well. This is apparent in the second frame at t = 15 units. This central G-well generates a region that is sufficiently supercritical to allow the fluctuation to rapidly grow in size and eventually develop into an autonomous particulate dissipative structure which is seen fully developed in the last frame at t = 35 units.

The particle shown here would represent a neutron. Its electric field consists of a Gaussian central core of high-Y/low-X polarity surrounded by a pattern of concentric spherical shells where X and Y alternate between high and low extrema of progressively declining amplitude. Since it is a reaction-diffusion wave pattern, we may appropriately name this periodicity the particle's Turing wave (LaViolette 2008b). The antineutron would have the opposite polarity, high-X/low-Y centered on a G potential hill.

The positive Y potential field (negative X potential field) in the neutron's core corresponds to the existence of a positive electric charge

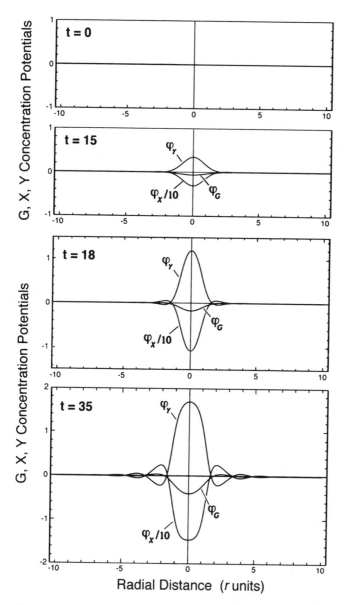

Figure 1.5. Sequential frames from a three-dimensional computer simulation of Model G showing the emergence of an autonomous dissipative structure particle: t = 0; the initial steady state; t = 15; growth of the positively charged core as the X seed fluctuation fades; t = 18; deployment of the periodic electric field Turing wave pattern; and t = 35; the mature dissipative structure particle maintaining its own supercritical core G-well. (Simulation by M. Pulver.)

density, and the surrounding shell pattern, which alternates between low and high Y potentials, constitutes shells of alternating negative and positive charge density. On the average, however, these charge densities cancel out to zero in the case of the neutron, which is why the Turing wave for the simulated neutron shown in figure 1.5 has no positive or negative bias with respect to the ambient zero potential.

The appearance of these positive and negative charge densities necessitates the simultaneous appearance of the particle's inertial rest mass. The shorter the wavelength of the Turing wave, and greater its amplitude (greater its etheron concentration wave amplitude), the greater will be the inertial mass of the associated particle (LaViolette 1985b). Since acceleration requires a structural shift and recreation of the particle's Turing-wave dissipative space structure, the particle's resistance to acceleration, its inertia, should be proportional to the magnitude of its Turing-wave charge densities; that is, proportional to the amount of negentropy that must be restructured (LaViolette 2013).

Subquantum kinetics further requires that for Model G to be physically realistic, the values of its kinetic constants, diffusion coefficients, and reactant concentrations should be chosen such that the emergent Turing wave has a wavelength equal to the Compton wavelength, λ_0, of the particle it represents, this being related to particle rest mass energy E_0, or to its rest mass m_0, by the formula:

$$(4) \quad \lambda_0 = hc/E_0 = h/m_0c$$

Here, h is Planck's constant and c is the velocity of light. The Compton wavelength for the nucleon calculates to be 1.32 fermis ($\lambda_0 = 1.32 \times 10^{-13}$ cm). The prediction that a particle's core electric field should have a Compton wavelength periodicity has since been confirmed by particle scattering experiments. Moreover, unlike the Schrödinger linear wave packet representation of the particle, which has the unfortunate tendency to progressively dissipate over time, the localized dissipative structures predicted by Model G maintain their coherence, as the underlying ether reaction-diffusion processes continuously combat the increase of entropy. Thus the Schrödinger wave equation used in quantum mechanics offers a rather naive linear approximation to representing microphysical phenomena, the quantum level being better described by a nonlinear equation system such as Model G.

Since this Turing wave particle representation incorporates both particle and wave aspects, we are able to dispense with the need to adopt a wave-particle dualism view of quantum interactions. Moreover the Turing wave subatomic particle has been shown to quantitatively account for the results of particle diffraction experiments, thereby eliminating paradoxes that arise in standard theories that rely on de Broglie's pilot wave interpretation of Schrödinger's wave packet concept. It also correctly yields Bohr's orbital quantization formula for the hydrogen atom while at the same time predicting a particle wavelength for the ground state orbital electron that is ~1,400 times smaller than Schrödinger's wave packet prediction. This more compact representation of the electron allows the existence of smaller diameter subground state orbits having fractional quantum numbers. Several researchers, such as John Eccles and Randall Mills, claim to have developed methods of inducing electron transitions to such subground orbits and to thereby extract enormous quantities of energy from plain water. Reformulating quantum mechanics on the basis of the subquantum kinetics Turing wave concept opens the door to understanding and developing new environmentally safe technologies.

In the course of dispensing with the Schrödinger wave packet and its associated probability function describing the indeterminate position of a mass point, it is advisable to also throw out the Copenhagen interpretation with its mysterious "collapse of the wave function" theorized to take place when the quantum "entity" through measurement becomes determined to be either a wave or a particle. In particular, Dewdney et al. (1985) have shown experimentally that the position of the particle is defined in a real sense prior to its de Broglie scattering event and from this conclude that in this particular case the wave-packet-collapse concept is flawed. More than likely, we should also be able to avoid using this collapse concept in experiments observing the spin orientation of entangled particles or polarization of entangled photons. There appears to be the growing realization that its widespread use is mainly a convenient mechanism to cover up the fact that physicists currently have a poor understanding of the workings of the subquantum realm.

When a neutron spontaneously acquires *positive charge* and transforms into a proton, its X-Y wave pattern acquires a positive bias similar to that shown in figure 1.6 (shaded region in the left hand plot). Such a biasing phenomenon, which is seen in analysis of the Brusselator, is also present in Model G when an existing ordered state undergoes a *secondary bifurcation*. The transition of the neutron to the positively biased proton state is best understood by reference to a bifurcation diagram similar to those used to represent the appearance of ordered states in nonequilibrium chemical reaction systems, as shown in figure 1.7.

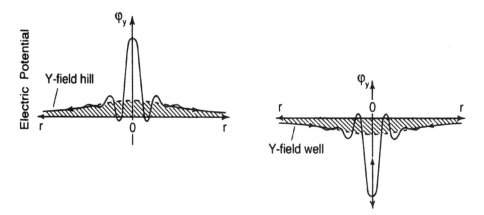

Figure 1.6. Radial electrostatic potential profiles for a proton and antiproton, positively charged matter state (left) and negatively charged antimatter state (right). The characteristic wavelength equals the particle's Compton wavelength.

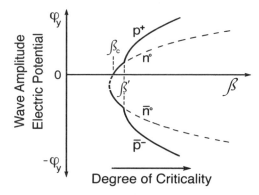

Figure 1.7. A hypothetical bifurcation diagram for the formation of nuclear particles. The secondary bifurcation past bifurcation point ß' creates electrostatic charge.

The emergence of the neutron from the vacuum state is represented as a transition to the upper primary bifurcation branch, which branches past critical threshold $ß_c$. Past threshold ß' this primary branch undergoes a secondary bifurcation with the emergence of the proton solution branch. This transition is observed in the phenomenon of beta decay, which also involves the production of electron and antineutrino particles not charted here; that is, $n \rightarrow p + e^- + \bar{\nu}^\circ + \gamma$.

The neutron's transition to the charged proton state involves an excess production rate of Y per unit volume in its core coupled with a corresponding excess consumption rate of X per unit volume. This causes a positive bias in its central Y concentration and a negative bias in its central X concentration, which in turn extends radially outward to bias the particle's entire Turing wave pattern. This extended field bias constitutes the particle's long-range electric field. Analysis shows that this potential bias declines as the inverse of radial distance, just as classical theory predicts. In fact, subquantum kinetics has been shown to reproduce all the classical laws of electrostatics, as well as all the classical laws of gravitation.

It should be kept in mind that the charge densities forming the proton's Turing wave pattern, and that are associated with its inertial mass, are distinct from and additional to the charge density that centrally biases its Turing pattern and produces the particle's long-range electric field. The former periodic densities emerge as a result of the particle's primary bifurcation from the homogeneous steady-state solution, while the latter aperiodic bias emerges as a result of its secondary bifurcation from an existing steady-state Turing solution.

Based on the results of the Sherwin-Rawcliffe experiment (Phipps 2009), we may infer that the creation and later displacement of the particle's Turing wave field would be communicated outward essentially instantaneously or at an exceedingly high superluminal velocity. The same would hold for the outward moving event boundaries of a sub-atomic particle's long-range electric and gravity potential fields. For their experiment, Sherwin and Rawcliffe (1960) performed mass spectrometry measurements of a football-shaped Lu^{175} nucleus to check for the presence of line-splitting and came up with a null result. This indicated that the mass of the lutetium nucleus behaved as a scalar instead of a tensor, which implied that its Coulomb field moved rigidly with its nucleus and was thereby able to create instant action-at-a-distance. Accordingly, the conventional practice of retarding force actions at speed c would be inappropriate.

Subquantum kinetics leads to a novel understanding of force, acceleration, and motion. In subquantum kinetics the energy potential field (ether concentration gradient) is regarded as the real existent and the prime cause of motion, "force" being regarded as a *derived* manifestation; that is, force is interpreted as the *stress effect* that the potential gradient produces on the material particle due to the distortion it manifests on the field pattern space structure that comprises the particle. The particle relieves this stress by homeostatic adjustment, which results in a jump acceleration and relative motion.

PARTICLE SCATTERING CONFIRMATION

The Turing wave configuration of the nucleon's electric potential field predicted by subquantum kinetics has been confirmed by particle scattering experiments that employ the recoil-polarization technique. Kelly (2002) has obtained a good fit to particle scattering form factor data by representing the radial variation of charge and magnetization density with a relativistic Laguerre-Gaussian expansion; see figures 1.8(a) and 1.9(a). The periodic character of this fit is more apparent when surface charge density $(r^2\rho)$ is plotted as a function of radial distance as shown in figures 1.8(b) and 1.9(b). Kelly's charge density model predicts that the proton and neutron both have a Gaussian shaped positive charge density core surrounded by a periodic electric field having a wavelength approximating the Compton wavelength. Moreover he has noted that unless this surrounding periodicity is included, his nucleon charge and magnetization density models do not make a good a fit to form factor data.

Thus here we have a stunning confirmation of a central feature of the subquantum kinetics physics methodology, whose prediction was first made in the mid-1970s at a time when it was still convention to regard the field in the core of the nucleon as rising sharply to a central cusp. Note also that Kelly's model confirms the positive biasing of the proton's central field, the bias increasing as the center of the

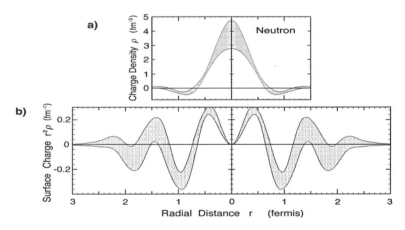

Figure 1.8. (a) Charge density profile for the neutron predicted by Kelly's preferred Laguerre-Gaussian expansion models and (b) the corresponding surface charge density profile (after Kelly 2002, Fig. 5–7, 18).

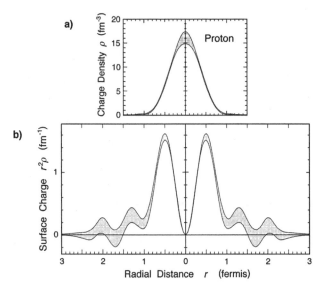

Figure 1.9. (a) Charge density profile for the proton predicted by Kelly's preferred Laguerre-Gaussian expansion models and (b) the corresponding surface charge density profile (after Kelly 2002, Fig. 5–7, 18).

particle is approached (compare the enhanced view shown in figure 1.9(b) with figure 1.6). Furthermore, as in the subquantum kinetics model, Kelly's model shows the amplitude of the nucleon's peripheral periodicity declining with increasing radial distance.

Simulations performed on Model G show that the amplitude of the Turing wave pattern declines with increasing radial distance as $1/r^4$ at small radii ($r < 2\lambda_0$), which approximates the radial decline observed in the charge density maxima for Kelly's model. The Model G particle Turing wave pattern declines more steeply at greater radial distances, declining as $1/r^7$ at $r \approx 4\lambda_0$ and $1/r^{10}$ for $r \approx 6\lambda_0$, which may be compared to standard theory, which proposes that the nuclear force declines as $F_n \propto 1/r^7$. This localized particle wave pattern is possible only because the extra G variable has been introduced into the Model G reaction system. It allows a particle to self-nucleate in an initially subcritical environment while leaving distant regions of space in their subcritical vacuum state. Thus if we quantify the amount of order or negentropy created by a single seed fluctuation, integrate the total amount of field potential $|\varphi_x|$ or $|\varphi_y|$ forming the particle wave pattern, we should find that it converges to a finite value, comparable to the idea of a quantum of action. The two-variable Brusselator, on the other hand, fails to generate localized structures. Simulations show that a seed fluctuation in the Brusselator produces order only if the system initially operates in the supercritical state, which in turn causes its entire reaction volume to become filled with a Turing wave pattern of maximum amplitude. Thus in the Brusselator a single seed fluctuation potentially produces an infinite amount of negentropy or structure.

The confirmation of the Model G ether reaction model, which has been forthcoming from particle scattering experiment data, leads us to conceive of the subatomic particle as an organized entity, or system, whose form is created through the active interplay of a plurality of particulate structures existing at a lower hierarchic level. Thus we find that the very structure of matter, its observationally confirmed Turing wave character, stands as proof of an underlying Whiteheadian dynamic and interactive stratum, one that ancient cultures variously named the Ether, Akasha, Tao, or Cosmic Ocean. The physics of subquantum kinetics has very ancient origins indeed (LaViolette 2004).

Contemporary quark models fail to anticipate the periodic character of the nucleon's electric field. No quark model can be devised after the fact that can reasonably account for this feature. Quarks themselves, or the "gluons" theorized to bind them together, have no script to tell them they should dance around in the complex manner that would be required in order to generate such an extended periodic field pattern. Subquantum kinetics, the viable replacement for quark theory, differs in several respects, one being the manner in which it handles the origin of mass, charge, and spin. Quark theory does not attempt to explain how inertial mass, electric charge, or spin arise. It merely assumes them to be physical attributes present in quarks in fractional form and which in triplicate summation appear as corresponding properties detectable in the nucleon. By comparison, etheron reactants of subquantum kinetics have no mass, charge, or spin. These are properties that are predicted to arise only at the quantum level, and which amazingly emerge as corollaries of the Model G reactions. Mass and spin, as properties of the subatomic particle, emerge at the time the particle first comes into being, and charge, as noted earlier, emerges as a secondary bifurcation of the primary Turing bifurcation. Rather than being merely assumed, all of these fundamental properties of matter come about in a comprehendable manner in the subquantum kinetics theory.

Paul A. LaViolette, Ph.D., is president of the Starburst Foundation, an interdisciplinary research institute, and holds advanced degrees in systems science and physics. The author of *Genesis of the Cosmos, Earth Under Fire, Decoding the Message of the Pulsars, Secrets of Antigravity Propulsion,* and *Subquantum Kinetics,* he lives in New York.

HYPOTHESIS 1 REFERENCES

Besant, A., and C. W. Leadbeater. 1919. *Occult Chemistry: Clairvoyant Observations on the Chemical Elements.* London: Theosophical Publishing House.

Cornille, P. 1998. "Making a Trouton-Noble Experiment Succeed." *Galilean Electrodynamics* 9: 33.

Dewdney, C., et al. 1985. *Foundations of Physics* 15: 1031–42.

Feynman, R. P., R. B. Leighton, and M. Sands. 1964. Vol. II of *The Feynman Lectures on Physics.* Reading, Mass.: Addison-Wesley.

Glansdorff, P., and I. Prigogine. 1971. *Thermodynamic Theory of Structure, Stability, and Fluctuation.* New York: Wiley.

Graneau, N. 1983. "First Indication of Ampere Tension in Solid Electric Conductors." *Physics Letters* 97A: 253–55.

Kelly, J. 2002. "Nucleon Charge and Magnetization Densities from Sachs Form Factors." *Physical Review* C 66, no. 6, ID: 065203. Eprint: http://arXiv.org/abs/hep-ph/0204239.

Lafforgue, J.-C. 1991. *Isolated Systems Self-propelled by Electrostatic Forces.* French patent No. 2651388.

LaViolette, P. A. 1985a. "An Introduction to Subquantum Kinetics: I. An

Overview of the Methodology." *International Journal of General Systems* 11: 281–93.

———. 1985b. "An Introduction to Subquantum Kinetics: II. An Open Systems Description of Particles and Fields." *International Journal of General Systems* 11: 305–28.

———. 1985c. "An Introduction to Subquantum Kinetics: III. The Cosmology of Subquantum Kinetics." *International Journal of General Systems* 11: 329–45.

———. 1994. *Subquantum Kinetics: The Alchemy of Creation*. Alexandria, Va.: Starlane Publications, first edition (out of print).

———. 2004. *Genesis of the Cosmos*. Rochester, Vt., Bear & Co. First ed. 1995.

———. 2008a. *Secrets of Antigravity Propulsion*. Rochester, Vt.: Bear & Co.

———. 2008b. *International Journal of General Systems* 37, no. 6: 649–76.

———. 2013. *Subquantum Kinetics: A Systems Approach to Physics and Astronomy*. 4th ed. Niskayuna, N.Y.: Starlane Publications.

Lefever, R. 1968. "Dissipative Structures in Chemical Systems." *Journal of Chemical Physics* 49: 4977–78.

Nicolis, G., and I. Prigogine. 1977. *Self-organization in Nonequilibrium Systems*. New York: Wiley-Interscience.

Pappas, P. T., and T. Vaughan. 1990. "Forces on a Stigma Antenna." *Physics Essays* 3: 211–16.

Phipps, Jr., T. E. 2009. "The Sherwin-Rawcliffe Experiment—Evidence for Instant Action-at-a-distance." *Apeiron* 16: 503–15.

Podkletnov, E., and G. Modanese. 2011. "Study of Light Interaction with Gravity Impulses and Measurements of the Speed of Gravity Impulses." In: *Gravity-Superconductors Interactions: Theory and Experiment*. Edited by G. Modanese and R. Robertson. Bussum, Netherlands: Bentham Science Publishers.

Prigogine, I., G. Nicolis, and A. Babloyantz. 1972. "Thermo-dynamics of Evolution." *Physics Today* 25, no. 11: 23–28; 25, no. 12: 38–44.

Pulver, M., and P. A. LaViolette. 2013. "Stationary Dissipative Solutions of Model G." *International Journal of General Systems* 42, no. 5: 519–41.

Sagnac, G. 1913. "The Luminiferous Ether Demonstrated by the Effect of the Relative Motion of the Ether in an Interferometer in Uniform Rotation." *Comptes Rendus de l'Academie des Sciences* (Paris) 157: 708–10.

Sherwin, C. W., and R. D. Rawcliffe. 1960. Report I-92 of March 14, 1960, of the Consolidated Science Laboratory (University of Illinois, Urbana); Dept. of Commerce Scientific and Technical Information Service, document #625706.

Silvertooth, E. W. 1987. "Experimental Detection of the Ether." *Speculations in Science and Technology* 10: 3–7.

———. 1989. "Motion through the Ether." *Electronics and Wireless World*: 437–38.

Winfree, A. T. 1974. "Rotating Chemical Reactions." *Scientific American* 230: 82–95.

Zaikin, A., and A. Zhabotinskii. 1970. "Concentration Wave Propagation in Two-dimensional Liquid-phase Self-oscillating System." *Nature* 225: 535–37.

HYPOTHESIS 2:
THE UNIVERSAL QUANTUM FIELD

Peter Jakubowski

This study gives a quantitative demonstration that the whole of contemporary physics, with all its equations, units, and constants, can be redefined and unified on the basis of the Universal Quantum Field, an interpretation of the Akasha dimension of the cosmos. The UQF is observable only through its quantized fluctuations, named fluctuons, which move across the field, exist for some time, then disappear. The manifest world is a composition of fluctuons.

The basic equations of contemporary physics emerge as relations among the "unified family" of the physical quantities generated by deduction from the Universal Quantum Field. The UQF does not need definition by observed and experimentally measured values beyond the Planck constant and the elementary electric charge. The unified family of all physical quantities "produces" all physical equations as simple relations between these quantities. The equations appear in their quantized, relativistic, and matter-independent form. —E.L.

INTRODUCTION

Our knowledge of the physical properties of the hidden dimension of the universe is just beginning to assume a solid place in contemporary science. This study shows that the physical properties of this Akasha dimension serve as the basis for describing all the physical phenomena in the universe.

In order to develop the new physics, we demonstrate how a purely physical Universal Quantum Field (UQF) can redefine and unify the whole of contemporary physics, with all its physical equations, units and constants, supported only by two "classical" values: the Planck constant h, and the elementary electric charge e.

We begin by describing the UQF. This field is observable only through its quantized fluctuations (named fluctuons), which move across the field, exist for some time, then disappear again. The observed world is but a composition of fluctuons.

This is correct, but it is not enough. We need a complete physics that can give us a description of the various physical quantities of the known world.

THE PHYSICS OF THE UQF

How can we characterize a fluctuon? Let us consider it as an object having its quantum wave vector **k** and quantum vector of propagation-velocity **c** (we note all vector physical quantities in bold letters). Let us call these the fundamental physical quantities, **k** and **c.** What is the relation of these quantities to each other? What other physical quantities can we define on the basis of **k** and **c**? And what other possible relations exist between the additionally defined physical quantities? Are all of them known in mainstream physics? In order to answer these questions let us introduce the following two-dimensional diagram (figure 2.1), which is a chessboard, with a place for every basic physical quantity we wish to define and correlate.

Figure 2.1 shows the central part of the Unified Family with the

R\C	-1	0	1
-1			**c** velocity
0	**k** wave vector	$^o1$ universal unity	
1			

Figure 2.1. Fundamental physical quantities, the vectors **k** and **c**, in relation to the scalar universal unity, in the central position of the Unified Family of all physical quantities.

quantum wave vector **k** and the quantum velocity **c** placed directly near the "universal unity" of the family. This unity-quantity plays an important role in the whole family, because many of the traditionally used physical quantities are in simple reciprocal relationship with each other; thus we need not define them separately. For example, we are able to define immediately the quantum length **r** as a vector reciprocal to the wave vector **k**, **r** = $(1/k)\hat{u}$, where $\hat{u}$ is a corresponding unit vector, and put it in the position opposite to **k**, on the right side of the universal unity, as shown in figure 2.2.

Our first question is: What is the relation of the fundamental physical quantities **k** and **c** to each other? The only such relation can be written as an outer product Λ of these two vectors. It defines the quantum frequency of the fluctuon as a bivector »f«. The bivector is an oriented plane, here a plane over **k** and **c** with a determined circulation along **k** and **c**. It is noted in figure 2.2 and in all following figures with a double line above the bivector quantity.

$$(1) \quad »f« = \mathbf{k} \wedge \mathbf{c}.$$

R\C	-1	0	1
-1		$\overline{\overline{f}}$ frequency	**c** velocity
0	**k** wave vector	$\overset{o}{1}$ universal unity	**r** length
1		$\overline{\overline{t}}$ period	

Figure 2.2. Central part of the Unified Family of all physical quantities. The bivector of the quantum frequency »f« is defined with Eq.(1) (see above), the quantum period »t« is its reciprocal, »t« = (1/f)»u«, where »u« is a unit bivector, and the vector length **r** is reciprocal to the wave vector **k**, **r** = $(1/k)\hat{u}$, where $\hat{u}$ is a unit vector.

Equation (1) is not only the first physical equation of the unified description of the observable world; it is in principle the single basic equation of the UQF, the Universal Quantum Field. It describes every single fluctuon of this field. Theoretically, nothing more is necessary for the description, because our observed world consists exclusively of fluctuons.

However, we want to know something more about the world. And if so, we need to answer also the second question: What other physical quantities can we define on the basis of **k** and **c**?

Two examples of such additional quantities are already present in figure 2.2: they are the reciprocals of the quantum wave vector **k** and the quantum frequency »f«. We have named the reciprocal of the wave vector **k** quantum length **r**. This is also the quantum size of the fluctuon. The reciprocal quantum frequency »f« has been called quantum period in figure 2.2 (and defined as »t«= (1/f)»u«, where »u« is a unit bivector), but it means also a quantum time, which is bivector, and not a scalar, as in conventional physics. It is our first drastic difference to the classic understanding of the observed world. In the unified quantized world we have not the classical, linear flow of time, from the past to the future. The universal quantum time has always the sense of a characteristic period of some corresponding quantum system, even if that system is a milliard (billion) light-years big. Quantum time does not flow, it circulates.

We now consider what further physical quantities we can define on the basis of **k** and **c**. We enlarge the plane of figure 2.2 on the right and thereby obtain figure 2.3.

As shown here, by twice repeating the multiplication of the universal unity with the quantum length **r**, we obtain the bivector quantum area »A« (»A« = $r_1 \wedge r_2$). Multiplying it with r again gives the quantum volume **V** of the fluctuon (**V** = »A«*·**r**, where * means an eigenproduct), which is equivalent to the quantum momentum **p**. Finally we multiply it with **r** and obtain the scalar quantum of action J, J = **p*****r** (= »A«*·»A«).

But how do we know that the fluctuon volume **V** is equivalent to

R\C	-1	0	1	2	3	4
-1		$\bar{\bar{f}}$ frequency	**c** velocity			
0	**k** wave vector	$\overset{\circ}{1}$ universal unity	**r** length	$\bar{\bar{A}}$ area	**p** momentum	$\overset{\circ}{J}$ action
1		$\bar{\bar{t}}$ period				

Figure 2.3. The consecutive multiplication of the universal unity with quantum length r results in bivector quantum area »A«, vector quantum volume V (equivalent to quantum momentum, p), and finally in the scalar quantum of action J, J = p*r (= »A«*»A«).

its momentum **p**? We need to put this to the test. If it works with the other members of the Unified Family, it is correct. And if we have chosen the right place for the momentum, then we also have the right place for action. This is because the classical and the quantum equivalence of action in regard to angular momentum is given above: J = **p*r**.

Let us test the position for linear and angular momentum. Consider what physical quantities could fill row -1 in figure 2.3. In light of Newton's classical physics, we can suppose that multiplying momentum **p** with frequency »f« would give us the correct position for the quantum vector of force **F**, **F** = »f«*p. Therefore we place force **F** on the intersection of column 3 with row -1.

We also know that force times length gives energy. If we place energy W[†] to the right of force (in column 4 of row -1), then we have to define it as the bivector »W« = **F** Λ **r**. Finally let us note that the physical quantity in column 2 in the same row must be the scalar quantum flux of frequency Φ_f, because it equals the (two-dimensional) quantum frequency »f« times the quantum area »A«, Φ_f = »A«*»f«.

[†]We have to use the symbol W for work to note energy, because in the Unified Family the standard symbol E must be left to denote electric field strength.

R\C	-1	0	1	2	3	4
-1		$\bar{\bar{f}}$ frequency	c velocity	$^{\circ}\Phi_f$ frequency flux	F force	$\bar{\bar{W}}$ energy
0	k wave vector	$^{\circ}1$ universal unity	r length	$\bar{\bar{A}}$ area	p momentum	$^{\circ}J$ action
1		$\bar{\bar{t}}$ period		$^{\circ}m$ mass		

Figure 2.4. The position we assumed for quantum mass
m is shown to be in proper relation to the already defined
physical quantities, p, J, F and »W«.

In order to complete the dynamic plane of the Unified Family, we
now need the correct position for quantum mass m. This is indicated
in figure 2.4.

A brief analysis of the known relations between quantities that are
already defined proves the correctness of the proposed position for scalar quantum mass m.

One step further along the diagonal direction between the rows and
columns in figure 2.4 leads us from the universal unity to quantum
velocity c. This suggests that every step along this or a parallel direction
means multiplication with the quantum velocity c. This gives the following relations: $m^*c = p$; $p \wedge c = »W«$, and $m^*(c_1 \wedge c_2) = »W«$.

We now ask, what do we know about the units of the here defined
physical quantities? If the quantum mass is correctly included in the
Unified Family, then we have the following definition of its unit, the
kilogram: $kg = m^2s$ (because of the relation in the row 1: $»t«^*»A« = »A«^*»t« = m$).

Let us test this definition. In classical physics the unit of force
should be kg^*m/s^2. Using our definition of kilogram, we can deduce
that $N = m^2sm/s^2 = m^3/s$. Due to the relation $F^*»t« = p$ (in column
3), we obtain as the unit of momentum the quantity m^3, and this is
also the unit of volume V. It follows that the quantum volume V and

quantum momentum **p** are equivalent physical quantities, and as such they occupy the same place in the Unified Family.

Now, the unit of energy is Joule: $J = Nm = kg^*m^2/s^2 = m^2s^*m^2/s^2 = m^4/s$. Our column 4 gives the relation for quantum action $J = »W«^*»t«$. It gives for the unit of action the combination m^4, the unit of volume multiplied by meter. Also the position for quantum action has been chosen correctly. This is important. It means that we can take the well-known value of the Planck constant ($J_u = h = 6.626076 \times 10^{-34}$ Js) and calculate with its fourth root the universal value of quantum size in the observable world: $r_u = 5.073575 \times 10^{-9}$m; ($r_u^4 = h$).

We have confirmed the correctness of the positions of all physical quantities in figure 2.4. Let us thus take the most important discoveries up to this point. First, the quantum frequency is a bivector; its flux Φ_f describes circulation, rotation, or spinning of a fluctuon. Second, the quantum time does not flow, it circulates; it always has a sense of some shorter or longer period. Third, the quantum of energy is not a classical scalar, it is a bivector. This means that a quantum of energy contains all the information that in traditional physics we had to add by means of the properties of the energy carrier, the photon. Every quantum of energy carries its own spin and spatial orientation. In Unified Physics we do not need both energy quanta and photons; we do not need photons at all, as the unified quanta of energy are their own energy carriers. Moreover the universal "building blocks" of the observed world are not atomic or molecular objects; they are nano-size fluctuons. Finally, we have discovered something distinctly promising. We no longer need the kilogram standard anymore. We are able to calculate the mass of a quantum object from its characteristic quantum period and its size (or from any two other of its physical quantities).

Let us now complete the Unified Family. First we complete its dynamic plane, as shown in figure 2.5 on page 176.

Figure 2.5 shows all the physical quantities that constitute the dynamic basis of the Unified Family of physical quantities. As we already know, some of these quantities are simply reciprocals of the

R\C	-3	-2	-1	0	1	2	3	4
-3					**G** gravity factor			
-2				$°f^2$ frequency square	**a** acceleration	$\bar{c}^2$ speed square	**f F** force in time	$°P$ power
-1				$\bar{f}$ frequency	**c** velocity	$°\Phi_f$ frequency flux	**F** force	$\bar{W}$ energy
0	**n** density	$\bar{\Lambda}$ quantum Laplacian	**k** wave vector	$°1$ universal unity	**r** length	$\bar{A}$ area	**p** momentum	$°J$ action
1		$°\sigma$ electric conductiv.	ρ_m mass density	$\bar{t}$ period	**km** linear mass dens.	$°m$ mass		
2		$\bar{\varepsilon}$ dielectric factor	**C** electr. capacitance	$°\Phi_\varepsilon$ optical area				

Figure 2.5. The most frequently used physical quantities belonging to the dynamic plane of the Unified Family.

other, as shown already in figure 2.4. Quantum Laplacian »Δ« is simply the reciprocal of quantum area »A« and quantum density **n** is reciprocal of the quantum volume **V** (alias momentum **p**). The quantum mass density ρ_m is the reciprocal of the quantum velocity **c** (because of the relations $\rho_m = (m/V)\hat{u}$ or $\rho_m{}^*V = m$); another discovery resulting from our definition of unified mass.

The position of the physical quantities that we call frequency square and speed square may be obvious as well. They are defined correspondingly as $f^2 = $ »f_1«*»f_2« and »c^2« = $c_1 \wedge c_2$. They are often used in traditional mathematical physics. At the moment it is enough to note that the frequency square has as its reciprocal the optical area (used in some specific optical equations), and the speed square is the reciprocal of the so-called dielectric factor »ε« (because of the well-known relation »ε«*»c^2« = 1). This relation means, however, that the dielectric factor, like its flux (the optical area) belongs to the dynamic plane of the Unified Family. Also the electric capacitance resulting from a multiplication of the dielectric factor with the quantum length **r**, C = »ε«*$\mathbf{r}$, belongs to this plane. Though the names "dielectric factor" and "electric capacitance" suggest something new, both these quantities have been traditionally defined as purely dynamic, and not as electrodynamic, quantities.

In order to properly introduce the electric conductivity in the Unified Family, we need to define its electrodynamic plane. We will construct it below. But first we need to deal with some important physical quantities in the upper part of the dynamic plane in figure 2.5.

It is relatively easy to find the right position for quantum acceleration **a**. It belongs to the position defined by a multiplication of the quantum velocity **c** with the quantum frequency »f«, or by the multiplication of the square of frequency with the quantum length **r**, **a** = »f«****c** = f^{2*}**r**. It is also clear that quantum power should lie above energy, because of its definition as the amount of energy "applied" or "used" during a specific period of time, P = »f«**»W«.

The other defining relations are P = **F*****c** = »f«****F*****r**. The latter defines also the position of the physical quantity **fF** (I have called it force in time, because it describes the change of the quantum force during the corresponding quantum period »t« (as the reciprocal of »f«). The units of acceleration (m²/s) and power (W = J/s) arise directly from the above definitions.

The last physical quantity at the top row of figure 2.5 is the next important finding of Unified Physics. This quantum-vector quantity is called the gravity factor **G**, because it plays the same role in the definition of the unified force as the gravity constant does in the Newtonian force of gravity. However, the unified gravity factor is not just a constant; it is an "ordinary" physical quantity that describes the temporal change of quantum acceleration, **G** = »f«****a**. Its unit is m/s³. The gravity factor becomes zero for all motion with constant acceleration. This discovery radically changes our understanding of the relation of gravity to antigravity.

Next, before we calculate the universal value of the gravity factor and the universal values of the other physical quantities, we have to construct the electrodynamic plane within which we locate them.

Because of known historical reasons, the development of the classical physics of magnetism and electricity occurred independently of each other and also independently of the other parts of physics, such as dynamics and kinematics. It would be thus quite understandable that the physical quantities introduced into each of these fields of physics can be defined independently of the other physical quantities. Fortunately for us, however, our scientific predecessors were great scientists who were intuitively aware of the unity of nature. Besides, they were also smart enough to understand not only their own field of physics, but the other fields as well. Thus the system of electrodynamic physical quantities and units, developed during the eighteenth and nineteenth centuries, is compact and self-consistent.

We can employ the multiplication rules we use in the dynamic plane to calculate all possible electrodynamic quantities; that is, use the UQF's two fundamental quantities, the quantum wave vector **k** and the quantum velocity **c**. The single problem that remains to be solved, using intuition, experience, and some luck, is to find out the proper numerical relationship between both planes of the Unified Family, the dynamic and the electrodynamic. This is mapped in figure 2.6.

When we compare figure 2.6 with figure 2.5, we find that the electrodynamic place of the universal unity has been occupied by another scalar physical quantity, namely magnetic induction B, which is equivalent to the planar density of the electric current j.

R\C	-2	-1	0	1	2	3	4
-1			$\overline{\overline{\text{B}}}$f induction in time	fH mag. field in time	^{0}E electric field	U electric potential	$\overline{\overline{\Phi}}_E$ electric flux
0	$\overline{\overline{\text{B}}}$A induction Laplacian	Bk mag. wave vector	$^0\overline{\text{B}}=^0$j magnetic induction	H magnetic field	$\overline{\overline{\text{i}}}$ electric current	Φ_H magnetic flux	$^0\tilde{\mu}$ mag. dipol moment
1		ρ_q charge density	$\overline{\text{D}}$ planar charge	kq linear charge	^{0}q electric charge	qr el. dipol moment	

Figure 2.6. All useful physical quantities belonging to the electrodynamic plane of the Unified Family. (The numbering of the columns and rows is the same as in figure 2.5.)

Like the universal unity, this scalar quantum physical quantity, B = j, is also a universal constant, independent of the material state (we define the matter dependence of physical qualities in figure 2.7). This is just a numerical factor that transforms the corresponding part of the dynamic plane into the electrodynamic plane. For example, comparing the actual positions of the magnetic field **H** (in figure 2.6) and the quantum length **r** (in figure 2.5) we can write directly **H** = B*r (or **H** = j*r). Similarly, the quantum electric current »i« equals: »i« = j*»A« and the planar charge (or electric induction) »D« is equal to »D« = B*»t« (better known as Maxwell's displacement current: j = »f«*»D«). As a result of our definition of the quantum mass, we have a new relation between quantum mass and electric charge: q = j*m, which gives us another way to find the numerical factor that connects the two planes. This factor is the quotient of the quantum electric charge in regard to the quantum mass of an arbitrary fluctuon, j = q/m. If we estimate the electric charge of a fluctuon, we immediately obtain its mass also. And the estimation of the quantum mass determines also the electric charge.

In order to complete the unification of all physical quantities we need to find the expression of the electric charge in reference to the fundamental quantities **k** and **c** (or through their derivatives »f« and **r**). I found that the square of the electric charge is a member of the dynamic plane. It can be located on the intersection of row 1 and column 3, next to the quantum mass: q^2 = m*r = »t«*r^3. This location is the origin of some interesting relations; for example, q^{2*}»f« = **p**, q^{2*}c = J, or $q^{2*}f^2$ = **F**. The last named is the indicator for the correct place to find the still "hidden" quantity q^2.

The traditional expression for the force generated by two wires carrying electric current is well known: **F** = i^2. In our quantum language it is expressed almost the same way: **F** = (q*»f«)^2. The only difference is that here the electric current is the bivector »i« and the classic current is defined as a vector quantity. As a consequence, the quantity for the square of the electric charge is a vector, whereas its classic expression treats it as a scalar. This relatively minute difference has obstructed the complete unification of physical quantities for all these years.

The electric charge is the second physical quantity (after the Planck constant) that has been often and precisely measured: it is $q_u = e = 1.602177 \times 10^{-19}$C. We thus take this quantity as the second universal value in the Unified Family. With a new definition of electric charge $q = (\gg t \ll {}^* r^3)^{1/2}$, we are now able to calculate the universal value of the second fundamental physical quantity. It is $t_u = e^2/r_u{}^3 = 1.965526 \times 10^{-13}$s, and $c_u = r_u/t_u = 2.581281 \times 10^4$ m/s. This value of the fundamental quantity c is a well-known quantity in physics when we realize that the quantum velocity of a given material is equivalent to its quantum electric resistance. The same value as c_u above but expressed in Ohms has been measured by Klaus von Klitzing in his investigation of the Hall effect. It earned him the Nobel Prize in physics in 1985.

The third possible plane of thermodynamic physical quantities should contain only one relevant item, namely temperature. Thus we treat temperature separately and simply add it to the Unified Family of physical quantities. The resulting Unified Family is given in the two tables of figure 2.7.

The defining relation is given in the left upper corner for each quantity. This definition includes the direct dependence on the material factor μ. Having extracted the material dependence (with different powers of μ) the remaining part of the definition can be expressed through the fundamental universal values (noted with the subscript "u"). The definition of the corresponding unit is given in the right upper corner and the universal value below the symbol of the given physical quantity. The dynamic plane is shown in figure 2.7a (note the shifted positions of G and n in relation to figure 2.5 for the picture-size reduction), and the electrodynamic plane is shown in figure 2.7b.

One new property of the Unified Family is demonstrated in the two parts of figure 2.7, in the left corner of each square. The unified definition of every physical quantity contains a corresponding material factor μ, which takes on the value 1 for the average state of the universal quantum field UQF, a value between 1 and 0 for all possible states of

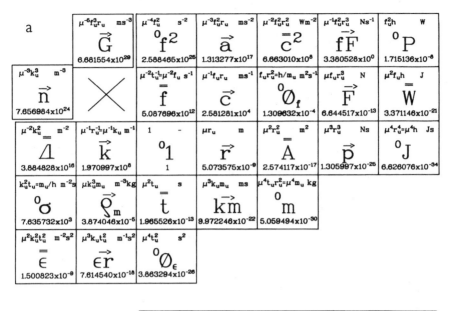

Figure 2.7. The two tables a and b of this figure offer the complete definition of the Unified Family of all physical quantities.

inanimate matter, and a value greater than 1 for all possible states of animate matter. (See the References on page 182 for a full definition of the quantum spectrum of matter and a detailed discussion of its properties.)

Let us return now to the Unified Family of physical quantities. We are speaking about all physical quantities, but figure 2.7 contains "only" 44 such quantities, or 45, including temperature. Where are the other hundreds known to science? The fact is that they are all equivalent to one or another of the chosen "basic" quantities shown in figure 2.7. For example, quantities equivalent to frequency flux, Φ_f, also include (in addition to the quantum of circulation and the quantum of rotation) the diffusion coefficient, elastic force coefficient, and electric resistivity. In addition to the already noted electric resistance, the energy density and pressure also belong to quantum velocity. The Unified Family does indeed define all physical quantities.

As a consequence of its completeness it also produces all possible relations between the quantities. We know them as physical equations, and some of them appear more important than others. Those we hold more important are known as the laws of physics, or even the laws of nature. This is misleading, because nature "works" without any physical equations. The only process I am ready to call a law of nature is the tendency in nature to transfer energy into the part of space where there is an energy deficit; or conversely, to transfer energy out of a region that has an energy excess in relation to its surroundings.

In science, and especially in physics, we have produced a great number of descriptions of phenomena and processes that seem unique and in need of a special explanation. The Unified Family offers explanations for 461 relations. We can read all of them directly from figure 2.7. We see, for example, that the Ohm equation for electric conductivity relates the electric field E from the electrodynamic plane with the electric conductivity from the dynamic plane. By comparing the place of the electric field with its corresponding physical quantity on the dynamic plane we find the scalar quantum flux of frequency Φ_f. On the other side, this frequency flux is reciprocal of the electric conductivity, $\Phi_f = 1/\sigma$. This gives the relation $E = j^*\Phi_f$, being the origin of the Ohm's relation. This means that we have properly chosen the "basic" place also of quantum electric conductivity on the dynamic plane.

In summary, we have shown that the Universal Quantum Field, constituting the Akasha dimension of the cosmos, does not need further reference to observed and experimentally measured values than the two (arbitrarily) chosen universal values, the Planck constant h and the elementary electric charge e. The Unified Family of all physical quantities, constructed in order to correctly define all basic quantities hitherto used in traditional physics, "produces" all physical equations as simple relations between these quantities. And it is noteworthy that these equations always appear in their quantized, relativistic, and matter-independent form.

In addition, UQF-based unification produces a number of quantitative relations unknown in mainstream physics, such as the relation between quantum mass and quantum electrical charge.

Peter Jakubowski earned his physics degree at Silesian University in Katowice, Poland, in 1969. His subsequent doctoral work from 1970 to 1973 focused on the electronic structure of matter. In addition to his theoretical investigations, Jakubowski completed an experimental study of the electronic structure of matter, earning his Ph.D. in 1976. He settled in Germany in 1984 and dedicated himself to developing a new, independent foundation for physics. This new physics paradigm, called Naturics, includes Unified Physics and its scientific and technological applications.

HYPOTHESIS 2 REFERENCES

Jakubowski, P. 1976. "Best Optimized One-electron Wave-functions; I. The General Procedure of Optimization; II. Isoelectronic Series of Li, Be, B, and C; III. Direct Examination of Optimization Effectiveness; IV. Ionization Energies of Atoms." *International Journal of Quantum Chemistry* X: 719–46.

———. 1978. "AFMR in KMnF3 near TN and TC; New Experimental Results." *Acta Physica Polonica* A54, no. 4: 397–409.

———. 1990a. "Luminiferous Ether Revived." *Physics Essays* 3, no. 3: 281–83.

———. 1990b. "Equivalence of Electrodynamics with Dynamics." *Physics Essays* 3, no. 2: 156–60.

———. 1992. "Alternative Foundation of Physics." *Physics Essays* 5, no. 1: 26–38.

———. 2003. *The Cosmic Carousel of Life: Our Evolution and Our Perspectives.* Norderstedt, Germany: Books on Demand GmbH.

———. 2010. *Naturics: The Unified Description of Nature.* Second edition. Norderstedt: Books on Demand GmbH.

References

Achterberg, J., K. Cooke, T. Richards, et al. 2005. "Evidence for Correlations between Distant Intentionality and Brain Function in Recipients: A Functional Magnetic Resonance Imaging Analysis." *Journal of Alternative and Complementary Medicine* 11, no. 6: 965–71.

Akimov, A. E., and G. I. Shipov. 1997. "Torsion Fields and Their Experimental Manifestations." *Journal of New Energy* 2: 2.

Akimov, A. E., and V. Ya. Tarasenko. 1992. "Models of Polarized States of the Physical Vacuum and Torsion Fields." *Soviet Physics Journal* 35: 3.

Arkani-Hamed, Nima, Jacob L. Bourjaily, Freddy Cachazo, et al. 2012. "Scattering Amplitudes and the Positive Grassmannian." DOI: .arXiv:1212.5605[hep-th].

Ashtekar, A., et al., ed. 2003. *Revisiting the Foundations of Relativistic Physics, Festschrift in Honor of John Stachel.* Boston Studies in Philosophy of Science. Vol. 234. Dordrecht, the Netherlands: Kluwer Academic.

Aspect, Alain, Jean Dalibard, and Gerard Roger. 1982. "Experimental Test of Bell's Inequalities Using Time-varying Analyzers." *Physical Review Letters* 49, no. 25: 1804–7.

Backster, Cleve. 1968. "Evidence of a Primary Perception at the Cellular Level in Plant Life." *International Journal of Parapsychology* 10, no. 4: 329.

Beloussov, Lev. 2002. "The Formative Powers of Developing Organisms." In *What Is Life?* Edited by Hans-Peter Dürr, Fritz-Albert Popp, and Wolfram Schommers. New Jersey, London, Singapore: World Scientific.

Bending, B. W. 2012. "Plant Sensitivity to Spontaneous Human Emotion." Poster session presented at: *Toward a Science of Consciousness.* April 10–14; Tucson, Ariz.

Biava, Pier Mario. 2009. *Cancer and the Search for Lost Meaning: the Discovery of a Revolutionary New Cancer Treatment.* Berkeley, Calif.: New Atlantic Books.

Byrd, Randolph. 1988. "Positive Therapeutic Effects of Intercessory Prayer in a Coronary Care Population." *Southern Medical Journal* 81: 7.

Chalmers, David J. 1995. "The Puzzle of Conscious Experience." *Scientific American* 273: 80–86.

Corichi, Alejandro. 2007. *Black Holes in Loop Quantum Gravity.* Morelia, Mexico: ICGC'07, IUCAA.

Del Giudice, Emilio, and R. M. Pulselli. 2010. "Structure of Liquid Water Based on QFT." *International Journal of Design & Nature and Ecodynamics* 5, no. 1.

Dossey, Larry. 1989. *Recovering the Soul: A Scientific and Spiritual Search.* New York: Bantam.

Dürr, S., T. Nonn, and G. Rempe. 1998. "Origin of Quantum-mechanical Complementarity Probed by a 'Which-way' Experiment in an Atom Interferometer." *Nature* 395, no. 3: 33–36.

Engel, Gregory S., Tessa R. Calhoun, Elizabeth L. Read, et al. 2007. "Evidence for Wavelike Energy Transfer through Quantum Coherence in Photosynthetic Systems." *Nature* 446 (12 April): 782–86.

Fodor, Jerry A. 1992. "The Big Idea." *New York Times Literary Supplement* (3 July).

Frecska, Ede, and Luis Eduardo Luna. 2006. "Neuro-ontological Interpretation of Spiritual Experiences." *Neuropsycho-pharmacologia Hungarica* 8, no. 3: 143–53.

Grof, Stanislav. 2012. "Revision and Re-enchantment of Psychology: Legacy of Half a Century of Consciousness Research." *The Journal of Transpersonal Psychology* 44, no. 2: 137–63.

Guth, Alan H. 1997. *The Inflationary Universe: The Quest for a New Theory of Cosmic Origins.* New York: Basic Books.

Hameroff, Stuart. 1987. *Ultimate Computing.* Amsterdam: North Holland Publishers.

Hameroff, Stuart, Roger Penrose, et al. 2011. *Consciousness and the Universe: Quantum Physics, Evolution, Brain & Mind.* Cosmology Science Publishers.

Hanada, Masanori, Yoshifumi Hyakutake, Goro Ishiki, and Jun Nishimura. "Description of Quantum Black Hole on a Computer." 21 November 2013, http://arxiv.org/abs/1311.5607. Accessed January 9, 2014.

Hawking, Stephen. 1974. "Black Hole Explosions?" *Nature* 248: 30–31.

Hawking, Stephen, and Leonard Mlodinow. 2010. *The Grand Design*. New York: Bantam.

Hoyle, Fred. 1983. *The Intelligent Universe*. London: Michael Joseph.

Hoyle, F., G. Burbidge, and J. V. Narlikar. 1993. "A Quasi-steady State Cosmology Model with Creation of Matter." *The Astrophysical Journal* 410, no. 23: 437–57.

Kafatos, Menas, and Robert Nadeau. 1999. *The Non-local Universe: the New Physics and Matters of the Mind*. Oxford: Oxford University Press.

Kuhn, Thomas. 1962. *The Structure of Scientific Revolutions*. Chicago: University of Chicago Press.

Kwok, Sun. 2011. *Organic Matter in the Universe*. New York: Wiley.

Kwok, Sun, and Yong Zhang. 2011. "Astronomers Discover Complex Organic Matter Exists throughout the Universe." *Science Daily* October 26.

Lanza, Robert, with Bob Berman. 2009. *Biocentrism: How Life and Consciousness Are the Keys to Understanding the True Nature of the Universe*. Dallas, Tex.: BenBella Books, Inc.

Laszlo, Ervin, and Kingsley Dennis. 2013. *The Dawn of the Akashic Age*. Rochester, Vt.: Inner Traditions.

———, with Anthony Peake. 2014. *The Immortality Hypothesis*. Rochester, Vt.: Inner Traditions.

LeShan, Lawrence. 2009. *A New Science of the Paranormal*. Wheaton, Ill.: Quest Books.

Linde, Andrei. 1990. *Inflation and Quantum Cosmology*. Boston: Academic Press.

———. 2004. "Inflation, Quantum Cosmology and the Anthropic Principle." In *Science and Ultimate Reality: From Quantum to Cosmos, Honoring John A. Wheeler's 90th Birthday*. Edited by John Barrow, Paul C. W. Davies, and C. L. Harper Jr. Cambridge: Cambridge University Press.

Mandel, Leonard. 1991. *Physical Review Letters* 67, no. 3: 318–21.

Megidish, E., A. Halevy, T. Sachem, et al. 2013. "Entanglement between Photons That Have Never Coexisted." *Physical Review Lett*ers 110: 210403.

Merali, Zeeya. 2007. "The Universe Is a String-net Liquid." *New Scientist* 15 March. *See also* clarification by Xiao-Gang Wen at http://dao.mit.edu/~wen/NSart-wen.html. Accessed October 16, 2013.

Mitchell, Edgar. 1977. *Psychic Exploration. A Challenge for Science.* New York: G. P. Putnam.

Montecucco, Nitamo. 2000. *Cyber: La Visione Olistica.* Rome: Mediterranee.

Nichol, Lee, ed. 2003. *The Essential David Bohm.* New York: Routledge.

Penrose, Roger. 1996. *Shadows of the Mind: A Search for the Missing Science of Consciousness.* Oxford: Oxford University Press.

————. 2004. *The Road to Reality.* London: Vintage Books.

Prigogine, Ilya, J. Geheniau, E. Gunzig, et al. 1988. "Thermodynamics of Cosmological Matter Creation." *Proceedings of the National Academy of Sciences USA* 85, no 20: 7428–32.

"Psionic Medicine." 2000. *Journal of the Psionic Medical Society and the Institute of Psionic Medicine* XVI.

Sági, Maria. 1998. "Healing Through the QVI-Field." In David Loye, ed. *The Evolutionary Outrider: the Impact of the Human Agent on Evolution.* England: Adamantine Press Limited.

————. 2009. "Healing Over Space and Time." In Ervin Laszlo, *The Akashic Experience.* Rochester, Vt.: Inner Traditions.

Sarkadi, Dezső, and László Bodonyi. 1999. "Gravity between Commensurable Masses." Private Research Center of Fundamental Physics, *Magyar Energetika* 7: 2.

Schrödinger, Ervin. 1969. *What Is Life? And Mind and Matter.* London: Cambridge University Press.

Sheldrake, Rupert. 2012. *The Science Delusion.* London: Hodder and Stoughton Ltd.

Smoot, George, and Keay Davidson. 1994. *Wrinkles in Time.* New York: William Morrow & Company.

Steinhardt, Paul J., and Neil Turok. 2002. "A Cyclic Model of the Universe." *Science* 296: 1436–39.

Susskind, Leonard. 2006. *The Cosmic Landscape: String Theory and the Illusion of Intelligent Design.* New York: Little, Brown & Company.

Trnka, Jaroslav. "The Amplituhedron." 19 September 2013, www.staff.science. uu.nl/~tonge105/igst13/Trnka.pdf. Accessed October 16, 2013.

Vivekananda, Swami. 1982. *Raja Yoga.* Calcutta: Advaita Ashrama.

Wheeler, John A. 1984. "Bits, Quanta, Meaning." In *Problems of Theoretical Physics.* Edited by A. Giovanni, F. Mancini, and M. Marinaro. Salerno, It.: University of Salerno Press.

About the Author

Ervin Laszlo spent his childhood in Budapest. He was a celebrated child prodigy on the piano, with public appearances from the age of nine. After receiving a Grand Prize at the international music competition in Geneva, he was allowed to cross the Iron Curtain and begin an international concert career, first in Europe, then in America. At the request of Senator Claude Pepper of Florida, he was awarded U.S. citizenship prior to his twenty-first birthday by an Act of Congress.

Laszlo received the Sorbonne's highest degree, the Doctorat ès Lettres et Sciences Humaines in 1970. Shifting to the life of a scientist and humanist, he lectured at various U.S. universities, including Yale, Princeton, Northwestern, Houston, and the State University of New York. Following his work on modeling the future evolution of world order at Princeton, he was asked to produce a report for the Club of Rome, of which he was a member. In the late 1970s and early 1980s, Laszlo ran global projects at the United Nations Institute for Training and Research at the request of the Secretary-General. In the 1990s his research led him to the rediscovery of the Akashic field, which he has continued to study and expound ever since.

The author, coauthor or editor of 89 books that have appeared in a total of 24 languages, Ervin Laszlo has also written several hundred papers and articles in scientific journals and popular magazines. He is a member of numerous scientific bodies, including the International

Academy of Science, the World Academy of Art & Science, the International Academy of Philosophy of Science, and the International Medici Academy. He was elected member of the Hungarian Academy of Sciences in 2010.

Laszlo's autobiography was published in June 2011 under the title *Simply Genius! And Other Tales from My Life*. The recipient of various honors and awards, including honorary PhDs from the United States, Canada, Finland, and Hungary, Laszlo received the Goi Peace Award, the Japanese peace prize, in 2001 and the Assisi Mandir of Peace Prize in 2006, and was nominated for the Nobel Peace Prize in 2004 and 2005.

Ervin Laszlo is founder and president of The Club of Budapest and chair of the Ervin Laszlo Center for Advanced Study.

Index

BOOKS OF RELATED INTEREST

Science and the Akashic Field
An Integral Theory of Everything
by Ervin Laszlo

The Immortal Mind
Science and the Continuity of Consciousness beyond the Brain
by Ervin Laszlo with Anthony Peake

The Akashic Experience
Science and the Cosmic Memory Field
by Ervin Laszlo

Dawn of the Akashic Age
New Consciousness, Quantum Resonance, and the Future of the World
by Ervin Laszlo and Kingsley L. Dennis

The New Science and Spirituality Reader
Edited by Ervin Laszlo and Kingsley L. Dennis

New Consciousness for a New World
How to Thrive in Transitional Times and Participate in the Coming
Spiritual Renaissance
by Kingsley L. Dennis

Science and the Afterlife Experience
Evidence for the Immortality of Consciousness
by Chris Carter

Morphic Resonance
The Nature of Formative Causation
by Rupert Sheldrake

INNER TRADITIONS • BEAR & COMPANY
P.O. Box 388
Rochester, VT 05767
1-800-246-8648
www.InnerTraditions.com

Or contact your local bookseller